# Studies in Infrastructure and Control

The book series aims to publish top-quality state-of-the-art textbooks, research monographs, edited volumes and selected conference proceedings related to infrastructure, innovation, control, and related fields. Additionally, established and emerging applications related to applied areas like smart cities, internet of things, machine learning, artificial intelligence, etc., are developed and utilized in an effort to demonstrate recent innovations in infrastructure and the possible implications of control theory therein. The study also includes areas like transportation infrastructure, building infrastructure management and seismic vibration control, and also spans a gamut of areas from renewable energy infrastructure like solar parks, wind farms, biomass power plants and related technologies, to the associated policies and related innovations and control methodologies involved.

Nilesh Chothani · Dharmesh Patel · Chirag Parekh

# Advancement in Infrastructure and Design Technology of Power Transformer

Springer

Nilesh Chothani
Department of Electrical Engineering
School of Engineering and Technology
Pandit Deendayal Energy University
Gandhinagar, Gujarat, India

Dharmesh Patel
Department of Electrical Engineering
Government Engineering College
Bharuch, Gujarat, India

Chirag Parekh
Hyundai Power Transformers
Montgomery, AL, USA

ISSN 2730-6453 ISSN 2730-6461 (electronic)
Studies in Infrastructure and Control
ISBN 978-981-96-9419-8 ISBN 978-981-96-9420-4 (eBook)
https://doi.org/10.1007/978-981-96-9420-4



This Springer imprint is published by the registered company Springer Nature Singapore Pte Ltd.
The registered company address is: 152 Beach Road, #21-01/04 Gateway East, Singapore 189721, Singapore

# Preface

Transformers are fundamental components in the power system, bridging the gap between power generation and consumption by stepping voltage levels up or down with efficiency and precision. Despite their wide usage, the process behind designing a reliable and high-performance transformer is both intricate and methodical. The purpose of this book, *Advancement in Infrastructure and Design Technology of Power Transformer*, is to offer a structured and practical procedure to the design process of power transformer with emphasis placed on the conventional and modern approaches, respectively. Intended for engineering students, practicing professionals and researchers interested in the understanding of the theoretical and practical aspects of transformer design, this book is crafted. Progressively, each chapter builds upon the one prior, starting from the base level infrastructure and progressing to later advance computer-planned systems techniques. This book is a learning tool, as well as a design reference, and is also provided with the use of practical steps and systematized calculations of numerical examples. We expect this work may serve as an inspiration for a more precise, clearer, and more innovative way to design electrical transformers for the reader.

The entire book is organized into eight chapters as follows:

The foundational elements that make up the backbone of transformer design are presented in Chap. 1. It defines the infrastructure, tools, materials, and resources needed to perform a transformer engineering task. This chapter establishes the tone and background for the rest of the book, first setting the stage for why designers need design infrastructure but then moving into the technical aspects. Chapter 2 outlines the traditional methods used to design transformer cores and yokes. It explains the core types, material choices, flux densities, and geometric considerations that influence electromagnetic performance and efficiency. Chapter 3 presents a step-by-step methodology for determining the key dimensions of a transformer, including core diameter and winding design. The concepts are reinforced with detailed numerical examples to illustrate practical application. Chapter 4 focuses on the electrical characteristics of transformer windings. It discusses how to calculate the resistance and leakage reactance for both high-voltage (HV) and low-voltage (LV) windings, which directly affect the performance. This chapter also includes a numerical solution for the percentage regulation of the transformer. Chapter 5 presents the comprehensive design of the power transformer, including window, core, yoke, frame, number of turns, and height of LV and HV coils. Chapter 6 delves into performance analysis by calculating no-load current, core losses, copper losses, and estimating the total weight. It discusses how to optimize these parameters for maximum efficiency and cost-effectiveness. Once electrical and mechanical designs are complete, the next concern is heat management—a critical factor in transformer durability. Chapter 7 discusses various cooling techniques, such as ONAN, ONAF, and OFAF, and provides guidelines for sizing radiators and cooling ducts to maintain safe operating temperatures. Chapter 8 explores the application of software and computational methods in transformer design. This chapter includes MATLAB-based programs for optimizing transformer design techniques.

Gandhinagar, India — Nilesh Chothani
Bharuch, India — Dharmesh Patel
Montgomery, USA — Chirag Parekh

# Acknowledgments

The authors humbly express their heartfelt gratitude to the Almighty, whose blessings and guidance have made the completion and publication of this meaningful work possible. We are deeply indebted to our families, whose unwavering support, encouragement, and motivation have been instrumental throughout the journey of this book. Their belief in us has been a constant source of strength.

Our sincere appreciation goes to Dr. Dipankar Deb, Book Series Editor of *Studies in Infrastructure and Control* at Springer Nature, and Dr. Aninda Bose, Executive Editor at Springer Nature, for their invaluable support, guidance, and encouragement during the development of this book. We also extend our gratitude to Springer Nature and the entire production team for their dedication and care in the publication process.

We gratefully acknowledge the support extended by Pandit Deendayal Energy University (Gandhinagar), Government Engineering College (Bharuch), and Hyundai Power Transformers USA (Montgomery). Their consistent cooperation played a key role in facilitating the work presented in this book. We also wish to thank the faculty, staff, and personnel of these institutions for their continual assistance.

Finally, we extend our heartfelt thanks to everyone who, in one way or another, contributed to the successful completion of this book project.

Nilesh Chothani
Dharmesh Patel
Chirag Parekh

# Key Features of This Book

This book provides an overview of power transformer infrastructure and computer-aided design with numerical solutions. Currently, a variety of design methodologies are available to perform the transformer infrastructure perfectly in the power grid. However, there is still a significant amount of work to develop these techniques for transformer design, to improve both cost and efficiency. Therefore, we can design additional methods to enhance efficiency and decrease manufacturing costs. To ensure the reliability of the entire power system, it is necessary to examine various issues related to transformer design and explore potential solutions within the power transformer infrastructure. We have prepared the algorithms to follow a step-by-step procedure for transformer design. We validate the suggested algorithms on Mathematical Laboratory (MATLAB) software.

This book provides a comprehensive overview of both conventional and research advancements in power transformer design, along with numerical solutions. Innovative analysis techniques and evolutionary algorithms optimize the design of three-phase core-type power transformers. It is crucial to analyze and reduce stray losses, as they are related to the raw material. Any decrease in these losses will directly reduce the transformer's overall cost.

Researchers will also find this book useful, as well as those working in transformer design. The philosophy about the different computerized design schemes used for the transformer is covered in the book. The strategies for reducing loss and design cost of the transformer are also described. The researchers working on the design of power transformers can discover new algorithms and novel schemes of transformer design and the related content. In addition, researchers can develop innovative schemes from the content of the chapters in this book.

- *The book delivers the entire transformer design process from the foundational infrastructure to advanced computerized methods, ensuring that readers develop a complete understanding.*
- *Covers clearly defined procedures in a logical sequence starting at the core and winding stage and moving to the thermal management.*

- *Combines detailed numerical solutions and sample problems to facilitate bridging theory and practical application of the complex concepts.*
- *Highlights critical performance parameters optimization, such as losses, no-load current, and weight, encouraging efficient and cost-effective transformer design.*
- *It balances traditional design practices with modern software design-based approaches and with a preparation for both manual to automated design environments.*

# Contents


**1 Transformer Design Engineering and Infrastructure** .............. 1
1.1 Introduction .............................................. 1
1.2 Basics of Transformers .................................... 4
1.3 Types of Transformers ..................................... 5
1.3.1 Based on Application .................................. 6
1.3.2 Based on the Number of Phases ......................... 7
1.3.3 Based on Construction .................................. 9
1.3.4 Based on the Voltage Ratio ............................. 9
1.4 Different Parts of the Transformer ........................ 11
1.4.1 Winding ............................................... 11
1.4.2 Core .................................................. 14
1.4.3 Tank and Radiator ..................................... 15
1.4.4 Bushings .............................................. 16
1.4.5 Insulation Component .................................. 16
1.4.6 Tap Changing Device ................................... 18
1.5 Transformer Materials ..................................... 18
1.5.1 Copper ................................................ 19
1.5.2 Cold-Rolled Grain-Oriented (CRGO) Steel ............... 19
1.5.3 Solid Insulating Material ............................. 20
1.5.4 Insulating Oil ........................................ 21
1.5.5 Mild Steel ............................................ 21
1.6 Transformer Design Philosophy ............................. 22
1.7 Key Performance Parameters for Transformer Design and Development ......... 23
1.7.1 No-Load Losses ........................................ 24
1.7.2 Load Losses and % Impedance ........................... 25
1.7.3 Cooling Process ....................................... 25
1.8 Transformer Design Optimization (TDO) ..................... 26
1.8.1 Manufacturing Cost Minimization ....................... 26
1.8.2 Operating Cost Minimization ........................... 27

1.8.3 AI and ANN Techniques for TDO ... 27
1.9 Conclusion ... 28
References ... 28

**2 Design of the Core and Yoke Through Conventional Technique** ... 31
2.1 Introduction ... 31
2.2 Required Specifications of a Transformer Design ... 34
2.3 Design of Single-Phase and Three-Phase Transformers ... 34
2.3.1 Design of Single-Phase Transformer ... 34
2.3.2 Design of Single-Phase Core-Type Transformer ... 35
2.3.3 Design of Single-Phase Shell-Type Transformer ... 37
2.3.4 Design of Three-Phase Transformer ... 38
2.3.5 Design of Three-Phase Core-Type Transformer ... 39
2.3.6 Design of Three-Phase Shell-Type Transformer ... 41
2.4 Volts per Turn of the Winding ... 44
2.5 Section of Flux Density ... 45
2.6 Choice of Current Density ... 46
2.7 Essentials of Designing the Core ... 46
2.7.1 Square Core ... 48
2.7.2 Stepped Core ... 49
2.8 Window and Core Proportions ... 53
2.9 Design of Yoke ... 53
2.10 Conclusion ... 54
References ... 55

**3 Determination of Main Dimensions of the Transformer with Numerical Solutions** ... 57
3.1 Introduction ... 57
3.2 Computation of the Overall Dimensions ... 59
3.2.1 Single-Phase Core-Type Transformer ... 60
3.2.2 Three-Phase Core-Type Transformer ... 60
3.2.3 Single-Phase Shell-Type Transformer ... 61
3.2.4 Three-Phase Shell-Type Transformer ... 61
3.3 Design of Windings ... 63
3.4 Types and Choice of Winding ... 64
3.4.1 Cylindrical Winding ... 64
3.4.2 Helical Winding ... 65
3.4.3 Continuous Disk Winding ... 65
3.4.4 Layer Winding ... 66
3.5 Voltage Rating-Based Selection of Windings in Transformers ... 66
3.6 Numerical with Solutions to Obtain the Main Parameters ... 66
3.7 Conclusion ... 82
References ... 83

**4 Resistance and Leakage Reactance Estimation of HV/LV Winding** ... 87
4.1 Introduction ... 87
4.2 Resistance Calculation ... 89
4.3 Reactance Calculation ... 90
4.4 Evaluation of Leakage Reactance in a Core-Type Transformer Featuring Concentric Coils of the Same Length ... 91
4.5 Numerical with Solutions to Obtain Leakage Reactance and Percentage Regulation ... 93
4.6 Conclusion ... 95
References ... 95

**5 Overall Design of Power Transformer** ... 99
5.1 Introduction ... 99
5.2 Overall Design Process and Flowchart ... 101
5.3 Numerical Design Based on Given Data ... 104
5.4 Conclusion ... 112
References ... 112

**6 Effects of No-Load Current, Losses, and Frequency in Optimum Design of Transformer** ... 115
6.1 Introduction ... 115
6.2 No-Load Current in the Transformer ... 118
6.3 Magnetizing Current in Single-Phase Transformer ... 119
6.3.1 Core Loss Current ... 120
6.4 Magnetizing Current in Three-Phase Transformer ... 120
6.4.1 Magnetizing Volt-Ampere and Alternate Expression for Magnetizing Current ... 120
6.5 Numerical and Solutions ... 121
6.6 Transformer Losses ... 125
6.6.1 No-Load Losses ... 125
6.6.2 Calculation of Core Loss ... 126
6.6.3 Calculation of Copper Loss ... 126
6.6.4 Power Loss Formula ... 127
6.6.5 Total Copper Loss ... 127
6.7 Effects of Change in Frequency in Parameters of the Transformer ... 128
6.8 Optimum Design of Transformer ... 129
6.9 Conclusion ... 132
References ... 132

**7 Design of the Cooling System for Transformer** .................... 135
7.1 Introduction .................... 135
7.2 Types of Cooling in Transformers .................... 137
7.3 Temperature Rise and Design of the Cooling System .................... 140
7.4 Plain-Walled Tank's Temperature Rise .................... 140
7.5 Temperature Rise in the Tank with Tubes .................... 140
7.6 Design of Tank with Tubes .................... 141
7.7 Flowchart for Design of Cooling System .................... 143
7.8 Numerical on Design of Transformer Cooling .................... 143
7.9 Conclusion .................... 151
References .................... 152

**8 Computerized Design Methodology of Power Transformer** .................... 155
8.1 Introduction .................... 156
8.2 Computer-Based Transformer Design Methodology .................... 158
8.3 Steps in the Sequential Design Process of a Transformer .................... 160
8.3.1 No. of Turns Calculation .................... 160
8.3.2 Core Design .................... 162
8.3.3 Design of HV and LV Winding .................... 162
8.4 Calculation of Performance Parameter .................... 163
8.4.1 The Total Cost of the Active Part .................... 163
8.5 Core Step Calculation .................... 164
8.6 Winding Gradient and Stray Loss Calculation .................... 164
8.6.1 LV Winding Gradient .................... 164
8.6.2 HV Winding Gradient .................... 165
8.6.3 Stray Loss (LV) .................... 165
8.6.4 Stray Loss (HV) .................... 165
8.6.5 Current and Flux Density Calculation .................... 166
8.7 Ampere Turns (AT) Balance Calculation .................... 166
8.8 Capacitance Calculation .................... 167
8.8.1 Parallel Capacitance .................... 167
8.8.2 Series Capacitance .................... 167
8.9 Impulse Calculation .................... 168
8.10 Inrush Current Calculation .................... 168
8.11 Dynamic Short Circuit Withstand Capacity Calculation .................... 168
8.11.1 Computing the Asymmetrical Short Circuit Current's First Peak Value .................... 169
8.11.2 Asymmetrical Short Circuit Ampere Turns Calculation .................... 169
8.11.3 Hoop Stress .................... 169
8.12 Radial Bursting Force .................... 169
8.12.1 Internal Axial Compression .................... 170
8.12.2 Axial Imbalance Force Calculation Owing to Tapping in the Winding .................... 170
8.12.3 Maximum Compressive Pressure on Spacers .................... 170

8.12.4 Tensile Stress in Tie Rods/Flitch Plates ... 171
8.12.5 Bending Stress on Clamping Structure ... 171
8.13 Thermal Ability Calculation ... 171
8.14 Efficiency and Regulation Calculation ... 172
8.15 Tank Calculation ... 172
8.16 Pressure and Vacuum Calculation ... 172
8.17 Oil Calculation ... 173
8.17.1 Radiator and Cooling Calculation ... 173
8.18 Computer-Based Power Transformer Design ... 174
8.19 MATLAB-Based Program for Transformer Design Optimization ... 174
8.20 Conclusion ... 176
References ... 181

# About the Authors

**Dr. Nilesh Chothani** is working as an Assistant Professor in the Department of Electrical Engineering at Pandit Deendayal Energy University, Gandhinagar, India. He received his Ph.D. in Electrical Engineering from Sardar Patel University in 2013. He has 20 years of teaching experience at the UG/PG level. His research interests include Electrical Machines, Power System Analysis, Power System Protection, Power System Simulation and Modeling, Artificial Intelligence, Integration of Renewable Energy Sources to Grid, and Smart Grid. He has developed a state-of-the-art power system protection laboratory, including real-time operation of a digital/numerical relaying scheme. He has published 35 peer-reviewed refereed international journal papers, 7 national journal papers and 46 IEEE international conference papers. He has also published 6 books in the domain of power system protection and switchgear. He has successfully guided three Ph.D. candidates and is guiding three more Ph.D. research scholars. Three of his IEEE international conference papers were selected as best papers and awarded for work of excellence. He received the 'TECH GURU' Award from Gujarat Technological University (GTU) in 2021 for his outstanding contribution to the Engineering field. He has also completed a research grant funded by the Department of Science and Technology, New Delhi, India. He is also a lifetime member of ISTE, IE (India) and IEEE.

**Dr. Dharmesh Patel** is an Assistant Professor in the Department of Electrical Engineering at Government Engineering College, Bharuch, Gujarat, India. He received a B.E. degree from North Gujarat University, Patan, Gujarat in 1999 and Master's degree in power systems from the Sardar Patel University, Vallabh Vidhyanagar, Gujarat, India in 2002. He received a Ph.D. degree in electrical engineering from Sardar Vallabhbhai National Institute of Technology, Surat, India, in 2019. His field of research is power transformer design, analysis and protection. He has published more than 28 research articles in various international journals and IEEE conferences to date and 2 books with an international publisher. He is guiding three Ph.D. research scholars from Gujarat Technological University.

**Dr. Chirag Parekh** is working with Hyundai Power Transformers USA, a Global leader in Power transformer manufacturing. He has worked as Vice President—Technical for 20 years with Atlanta Electricals Pvt. Limited (AEPL), one of India's fastest-growing Power Transformer manufacturers. He received his Ph.D. in Electrical Engineering from RK University in 2017. His total industrial-academic experience is around 22 years. He has the most significant Power Transformer Design, Engineering and manufacturing expertise. He has developed innovative products like Mobile substation, Short circuit testing transformer, Ester oil-filled Power Transformer during his industrial tenure in India. He has also developed and established a fully automated and state-of-the-art Power Transformer Testing laboratory for testing power transformers up to 200 MVA capacity and 220 kV voltage class. He has published more than 20 papers in National and International journals and conferences. He has successfully guided one Ph.D. candidate. At present, he is a member of IEEE USA and CIGRE USA. He recently received an outstanding team member award from Hyundai Power Transformer USA.

# List of Figures


Fig. 1.1 Simplified layout of power system .......... 2
Fig. 1.2 Transformer situated in substation .......... 3
Fig. 1.3 Parts of a transformer .......... 4
Fig. 1.4 Classification of transformers .......... 5
Fig. 1.5 Three-winding transformer .......... 8
Fig. 1.6 Winding arrangements of three-winding transformer: **a** LV, MV, and HV; **b** MV, LV, and HV .......... 8
Fig. 1.7 Single-phase and three-phase core transformer .......... 9
Fig. 1.8 Single-phase and three-phase shell-type transformer .......... 10
Fig. 1.9 Step-up and step-down transformers .......... 11
Fig. 1.10 Different types of transformers. **a** Step-up and step-down transformer, **b** core- and shell-type transformer, **c** generator transformer, **d** power and distribution transformer, **e** single- and three-phase transformer, **f** furnace- and dry-type transformer, **g** traction and rectifier duty transformer, **h** testing transformer .......... 12
Fig. 1.11 Composite view of different parts of transformer [2] .......... 15
Fig. 1.12 Transformer winding and core assembly .......... 15
Fig. 1.13 Transformer tank and radiators .......... 16
Fig. 1.14 Transformer bushings (condenser and solid porcelain type) .......... 17
Fig. 1.15 Insulation components for transformer .......... 17
Fig. 1.16 Tap changer for transformer (OLTC and OCTC) .......... 18
Fig. 1.17 Copper and lamination .......... 19
Fig. 1.18 Cellulose base insulating material .......... 20
Fig. 1.19 Dielectric oil for power transformer .......... 21
Fig. 1.20 Mild steel for transformer tank fabrication .......... 22
Fig. 2.1 Cross-section of $1\varphi$ core-type transformer .......... 35
Fig. 2.2 Winding arrangement of $1\varphi$ core-type transformer .......... 36
Fig. 2.3 Cross-section of $1\varphi$ shell-type transformer .......... 38
Fig. 2.4 Winding arrangement of $1\varphi$ shell-type transformer .......... 38

Fig. 2.5 Cross-section of 3$\varphi$ core-type transformer .................. 39
Fig. 2.6 Winding arrangement of 3$\varphi$ core-type transformer ............ 39
Fig. 2.7 Cross-section of 3$\phi$ shell-type transformer .................. 41
Fig. 2.8 Winding arrangement of 3$\phi$ shell-type transformer ........... 42
Fig. 2.9 Shape of the various transformer cores and their laminations ... 47
Fig. 2.10 Types of cores ............................................. 47
Fig. 2.11 Core sections of **a** multi-stepped, **b** square, and **c** four-stepped cruciform cores ......................... 47
Fig. 2.12 Square core ................................................ 48
Fig. 2.13 Stepped core ............................................... 50
Fig. 2.14 Types of stepped cores ..................................... 52
Fig. 2.15 Shape of leg and yoke in the core assembly of a transformer ... 53
Fig. 3.1 Representations of overall dimensions in 1-∅ core-type transformer ................................................ 60
Fig. 3.2 Representations of overall dimensions in 3-∅ core-type transformer ................................................ 61
Fig. 3.3 Representations of overall dimensions in 1-∅ shell-type transformer ................................................ 62
Fig. 3.4 Representations of overall dimensions in 3-∅ shell-type transformer ................................................ 62
Fig. 3.5 Different types of windings used in transformers ............. 64
Fig. 3.6 Different types of windings used in transformers: **a** Continuous disk-type winding, **b** helical winding, **c** layer winding, and **d** cylindrical winding ....................... 65
Fig. 3.7 Window representation for Example 3.4 ...................... 73
Fig. 3.8 Main dimensions in window and core ........................ 76
Fig. 3.9 Main dimensions in window and core for Example 1.5.7 ...... 80
Fig. 5.1 Flowchart for overall design procedure ..................... 102
Fig. 5.2 Three-stepped core ......................................... 103
Fig. 5.3 Window design with dimensions ............................. 106
Fig. 6.1 Phasor diagram for no-load current ......................... 118
Fig. 7.1 Types of cooling system .................................... 137
Fig. 7.2 Cooling mechanisms: **a** OFAN, **b** OFAF, **c** ONAF, **d** ONAN with tubes, and **e** ONAN ......................... 138
Fig. 7.3 Transformer main dimensions with tank along with clearances ........................................ 144
Fig. 7.4 Flowchart for design of tank with tubes in transformer ........ 145
Fig. 7.5 Cooling tubes arrangement for example 7.1 .................. 146
Fig. 7.6 Cooling tubes arrangement for example 7.2 .................. 148
Fig. 7.7 Cooling tubes arrangement for example 7.3 .................. 150
Fig. 8.1 Transformer design process flowchart ...................... 160
Fig. 8.2 Parametric impact on the cost and losses of the transformer **a** NLL, load loss, and %Z comparison, **b** NLL versus cost of transformer, **c** load loss versus cost of transformer ......... 161

Fig. 8.3 Programming for design of power transformer using MS-excel ........ 175
Fig. 8.4 MATLAB flowchart for transformer design optimization process using iterative method ........ 177
Fig. 8.5 **a** Programming for design of power transformer using MATLAB and **b** respective result using iterative method ........ 178
Fig. 8.6 Graphical representation of iterative method TDO, **a** flux density versus NLL, **b** current density (HV) versus load loss, **c** current density (LV) versus load loss, **d** stiffener versus deflection in tank, and **e** *K*-factor versus no load loss ........ 180

# List of Tables

| | | |
|---|---|---|
| Table 1.1 | Functions of various parts of the transformer | 5 |
| Table 1.2 | Comparison between three-phase transformer bank and three-phase transformer | 7 |
| Table 1.3 | Comparison between core-type and shell-type transformers | 10 |
| Table 2.1 | Values of $K$ for various types of transformers | 45 |
| Table 2.2 | Values of flux density for various types of transformers | 45 |
| Table 2.3 | Values of current density for various types of transformers | 46 |
| Table 2.4 | ($A_{gi}/A_{cc}$) and ($A_i/A_{cc}$) for various cores | 51 |
| Table 3.1 | Voltage rating-based selection of windings in transformers | 67 |
| Table 6.1 | Summary of effects of frequency variation on various parameters | 129 |
| Table 7.1 | Types of heat transfer from various regions of transformer | 140 |
| Table 7.2 | Clearance values for a typical transformer | 143 |
| Table 8.1 | Variables, monitoring, and output parameters of transformer design | 159 |
| Table 8.2 | Design data of power transformer conventional design program by Ms-excel tool | 176 |
| Table 8.3 | List of variables for the TDO program | 178 |
| Table 8.4 | List of design constraints for the TDO program | 179 |

# Chapter 1
# Transformer Design Engineering and Infrastructure

Transformer engineering has a central role in guaranteeing the stability, dependability, and effectiveness of the present-day power frameworks. This chapter presents a detailed overview of transformer infrastructure, starting firstly, by commenting on the fundamental role of these instruments in electricity networks, and secondly, the different types of these instruments used at different voltage levels and in different applications. Multi-function devices and inelastic deformations under various constructive conditions are described; key performance parameters and transformer design scope are given, covering voltage regulation, efficiency, impedance, and losses determined under load and no-load conditions. The structure of transformers is broken up into its main parts (core materials, winding arrangements, and insulating systems containing oil types and their thermal properties). The chapter also offers to explore the cooling techniques of great importance to thermal management and provides the strategy of transformer design optimization. It is emphasized that manufacturing and operational costs and performance standards can be minimized. In addition, recent developments in artificial intelligence, such as artificial neural networks (ANNs), are introduced as powerful tools for optimization of transformer design. Using these techniques can enable increasing accuracy of modeling, increased design iteration speediness, and better predictions of performance. Overall, the chapter brings the engineering and computer intelligence together, building a progressive road to affordable, efficient, and intelligent transformer generation.

## 1.1 Introduction

In power systems, electricity generated at power plants must be transmitted effectively over long distances before reaching consumers. To minimize transmission losses, voltage levels must be raised to extra-high voltage (EHV) or ultra-high voltage (UHV) ranges. Modern transmission networks typically operate at voltages from 100



N. Chothani et al., *Advancement in Infrastructure and Design Technology of Power Transformer*, Studies in Infrastructure and Control,
https://doi.org/10.1007/978-981-96-9420-4_1

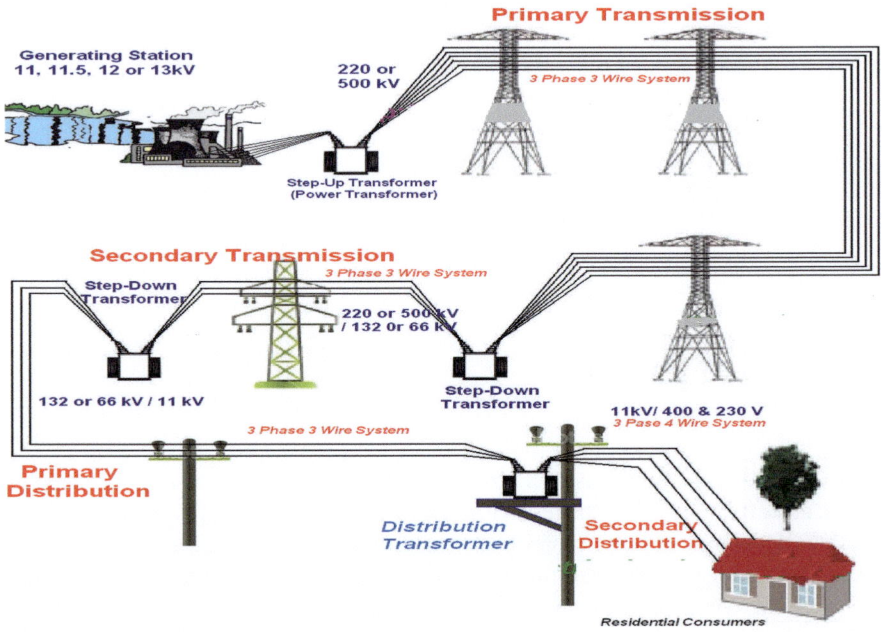


**Fig. 1.1** Simplified layout of power system

to 500 kV, with some lines reaching as high as 765 kV. However, these high-voltage levels are unsuitable for direct usage in residential, commercial, or industrial environments, necessitating a reduction in voltage through transformers. Furthermore, power generators are generally constructed to generate electricity at voltage levels between 10 and 40 kV AC to optimize cost and efficiency. As a result, transformers are necessary at the generating facility to increase the voltage to suitable transmission levels. Figure 1.1 depicts a simplified power system, highlighting the actual voltage levels present at various stages of transmission and distribution.

In power systems, it is usually essential to implement a series of voltage reductions from transmission to final distribution. This is achieved through the utilization of transformers. Figure 1.2 shows a transformer located in a switchyard. The high-voltage input line connects to the transformer's primary winding to lower the voltage for local use from the secondary winding of the transformer. The voltage from the secondary may be further decreased before it reaches the end consumer. A transformer can fulfill various roles, such as providing lighting or supplying electricity to homes, communities, or large industries. Despite significant advancements in technology, modern transformers still operate on the same fundamental principles. According to both national and international standards such as IEC, ANSI, and IEEE (IEEE Standard Terminology for Power and Distribution Transformers, IEEE Std C.57.12.80-2002) [1], a transformer is defined as a static electrical device made up of magnetic and electrical circuits, featuring two or more windings and a core that facilitates mutual coupling between electric circuits.

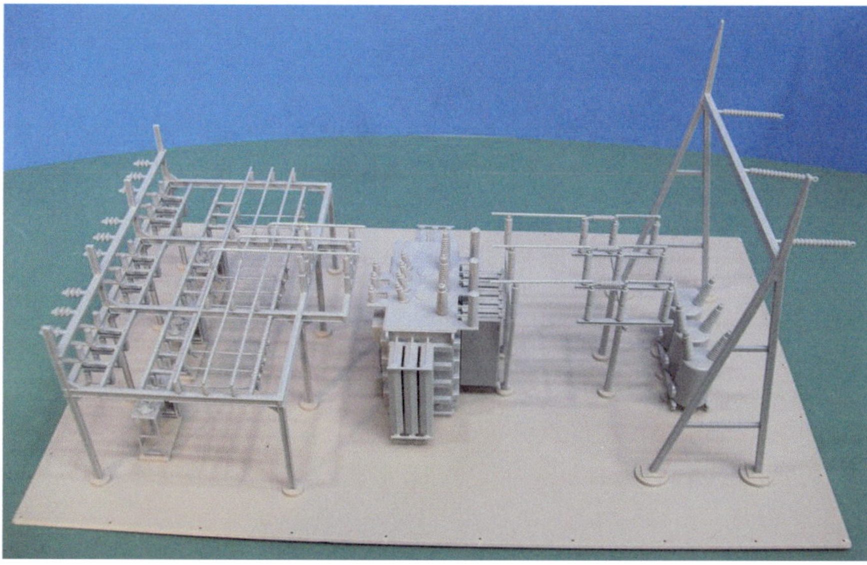

**Fig. 1.2** Transformer situated in substation

Given the intricate nature of interconnected systems, there is always the potential for either an increase or a decrease in voltage. As a result, it is often essential to make accurate voltage modifications to compensate for voltage losses in lines and other equipment. These voltage losses vary depending on the load current, leading to variations throughout the day. This is where tap-changing devices become essential. These devices modify the number of turns in a winding, thus altering its voltage. This adjustment can take place while the transformer is under load or when it is disconnected, known respectively as on-load or off-load tap changers. A solid understanding of electromagnetism, electrical circuit analysis, insulation coordination, magnetic circuit analysis, heat and mass transfer, and thermodynamics is crucial when designing a transformer. Transformers need design to determine component dimensions using information provided by customers to manufacturers. Researcher's completion of transformer design remains complicated because they need to meet operational requirements combined with cost reduction needs. An optimal design enables the manufacturer to deliver a high-performance product at a reduced cost. Enhancements to the design can be achieved by either reducing manufacturing expenses, overall ownership costs, costs associated with the transformer's active components, the weight of those active parts, or losses, or by maximizing the transformer's apparent power within given constraints.

## 1.2 Basics of Transformers

Transformers are static electromagnetic devices with two or more windings and a common magnetic circuit. They are one of the primary power apparatuses used in the power system network. The transformers change the AC voltage and the associated current from one level to another level, maintaining the power balance between the input and output sides, without a change in frequency. Hence, transformers perform AC energy transformation rather than energy conversion from one or more primary side circuit/s to one or more secondary side circuit/s.

The main parts of the transformer are as follows:

- Core.
- Yoke.
- Winding.

When the MVA rating is high, in addition to the above parts, the transformer is equipped with

- Bushing.
- Coolant conservator.

The different parts of the transformer in general are represented in Fig. 1.3.

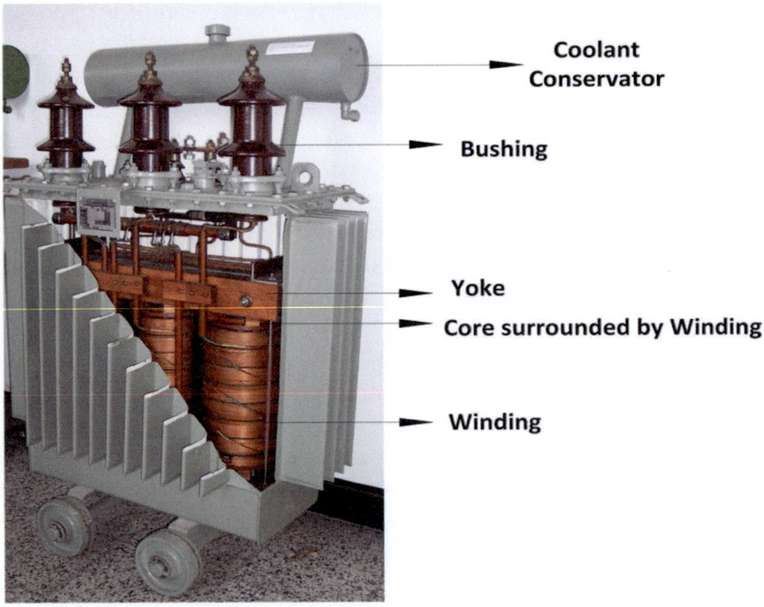


**Fig. 1.3** Parts of a transformer

## 1.3 Types of Transformers

The classification of transformers is represented in Fig. 1.4. In addition, instrument transformers such as current transformers and potential transformers are also available. The detailed classification based on the aforementioned aspects is given in the following section. The key functions and materials used for different parts of the transformer are stated in Table 1.1.

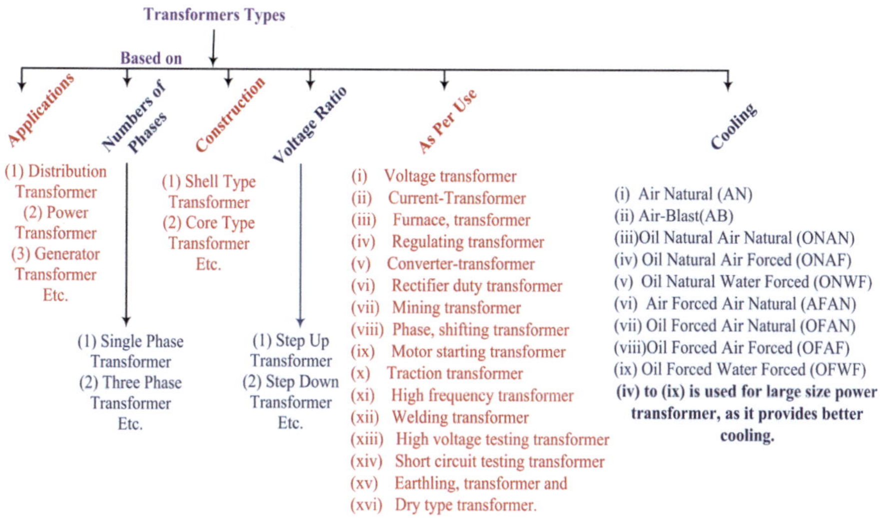


**Fig. 1.4** Classification of transformers

**Table 1.1** Functions of various parts of the transformer

| Part | Material | Function |
|---|---|---|
| Core | Cold-rolled grain-oriented steel (CRGO) | • Creation of a closed magnetic circuit for passage of flux between primary and secondary windings<br>• Windings are made around the core |
| Yoke | Cold-rolled grain-oriented steel (CRGO) | • It acts as the top and bottom in the core even if it also provides support to it |
| Winding | Copper | • Carries the current and is used for transfer of power |
| Bushing | Polymer or porcelain | • Prevention of flashover due to connection of high voltage to an earthing point |
| Coolant conservator | Steel | • Provides storage space for coolant used for circulation in cooling the transformer |

### 1.3.1 Based on Application

As far as concerns regarding power systems, there are three main categories involved as per the given standards.

(i) **Distribution transformers,** which have a rated power of up to 2.5 MVA and a voltage of up to 36 kV.
(ii) **Power transformers,** designed for a rated power of 1000 MVA and a voltage of 1500 kV.
(iii) **Generator transformers** are primarily used to connect the output of generators to the rest of the power system as step-up voltage transformers. It can also be defined based on construction.

- ***Distribution Transformer***: Such transformers are used for stepping down the transmission voltage to distribution voltage or stepping down the voltage level to rated voltage as per the service requirements of an industrial or commercial load. The rating of this transformer is usually below 1000 kVA. The type of connection used is a three-phase four-wire delta-star. The period of operation is 24 h a day. Hence, the design should be done in a way that the efficiency is maximum even at half-load conditions, leading to a good all-day efficiency. The flux density is usually less than 1.7 $Wb/m^2$. Here, 1:3 is the ratio of iron loss to copper loss. As this type of transformer is located in the vicinity of the load, the voltage regulation must be good, which can be satisfied by having a small leakage reactance. Usually, the percentage impedance is in the range of 4–18%. The voltage regulation is in the range of 4–9%. The type of cooling used is self-oil cooling.
- ***Power Transformer***: Such transformers are used to step up the voltage or step down the voltage in powerhouses and substations at both ends of the power transmission line. The rating of this transformer is usually above 1000 kVA. The type of connection used is three-phase three-wire-delta-delta or star-delta. The period of operation is during heavy loading periods and is disconnected during light loading periods. Hence, the design is in such a way that efficiency is maximum at full or near full load. The flux density is usually between 1.5 and 1.7 $Wb/m^2$. The ratio of iron loss to copper loss is 1:1. The voltage regulation is not an important factor, as this transformer is not located in the vicinity of the load. Hence, the leakage reactance can be made high and used to limit the short-circuit current, leading to less mechanical force. Usually, the percentage impedance is in the range of 6–18%. The voltage regulation is in the range of 6–10%. Here, forced air or forced oil cooling is used for cooling purposes.

### 1.3.2 *Based on the Number of Phases*

- ***Single-Phase Transformer***: Such of transformer has a single-phase primary and a single-phase secondary placed with either core or shell-type construction. It is used for low-power applications.
- ***Three-Phase Transformer***: This type of transformer has three-phase primary and three-phase secondary placed with either core or shell-type construction. It is used for high-power applications.

A three-phase transformer can also be constructed by combining three single-phase transformers into a star or delta connection, depending on the needs. We refer to this as a transformer bank with three phases. On the other hand, all three phases of a three-phase transformer can access shared magnetic circuits. Table 1.2 displays a comparison of the three-phase transformer and the three-phase transformer bank.

The three-$\phi$ transformer connections can be connected in star or delta type. In a star assembly, the phase voltage ($V_{\mathrm{ph}}$) $= \sqrt{1/3}$ times the line voltage ($V_{\mathrm{L}}$); so, the number of turns is $\sqrt{1/3}$ times the number of turns in the delta connection. But the line current ($I_{\mathrm{L}}$) is $\sqrt{3}$ times the phase current ($I_{\mathrm{ph}}$) in the delta connection, hence requiring a large area of cross-section. The amount of insulation used in the star connection is smaller compared to the delta connection. In star connection, neutral grounding/earthing can be done easily, reducing the effect of faults. For high voltages, the cost of a star connection is less.

Moreover, a four-wire connection with neutral (fourth wire) can be easily done. But, for the delta connection, under unbalanced conditions, the response is better than the star connection.

Apart from the above, there is another type of transformer with *three winding* (primary, secondary, and tertiary), as shown in Fig. 1.5 to operate at high, medium, and low voltages.

The reasons for the use of tertiary winding are as follows:

- Ability to supply an additional load with a small voltage difference between primary and secondary.
- To provide voltage for compensating capacitors for power factor correction at different voltages.

**Table 1.2** Comparison between three-phase transformer bank and three-phase transformer

| Feature/type | Three-phase transformer bank | Three-phase transformer |
|---|---|---|
| Cost | Costly, due to the large volume | Cheaper, due to small volume |
| Efficiency | Less | High |
| Installation and maintenance | Difficult | Easy |
| Faults and clearance | Easier to clean, as any single unit in which fault has occurred can be replaced | Difficult to clean, as it is a single unit |

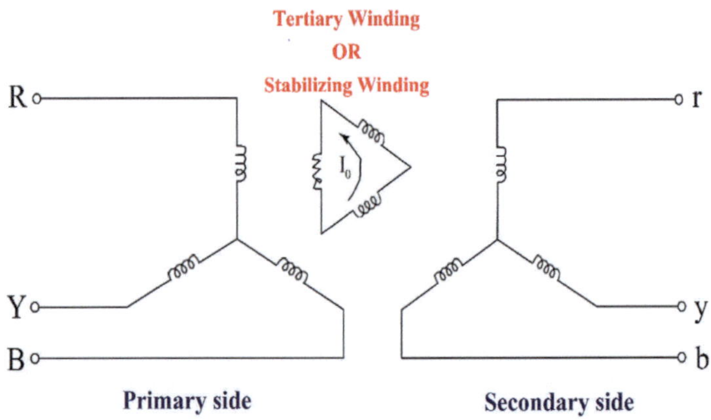


**Fig. 1.5** Three-winding transformer

- As a stabilizing winding for limiting short circuits or fault currents, when connected in a delta.
- For indicating voltage levels in testing transformer was used for high voltages.
- To reduce zero sequence impedance, making easy flow of fault current easy to ground.

The different possible arrangements of windings in three-winding transformers are shown in Fig. 1.6a and b.

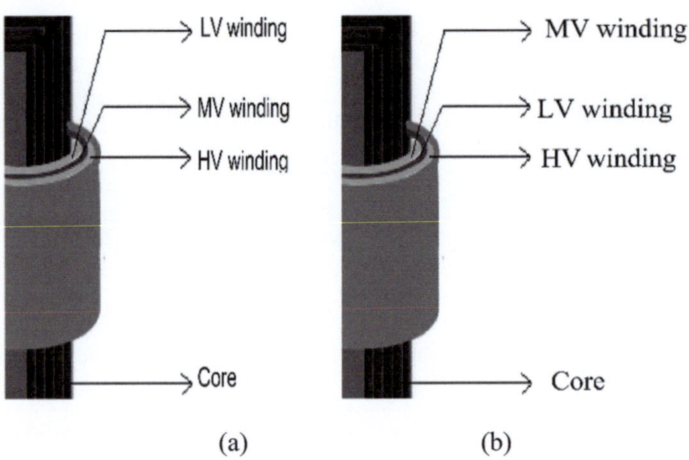


**Fig. 1.6** Winding arrangements of three-winding transformer: **a** LV, MV, and HV; **b** MV, LV, and HV

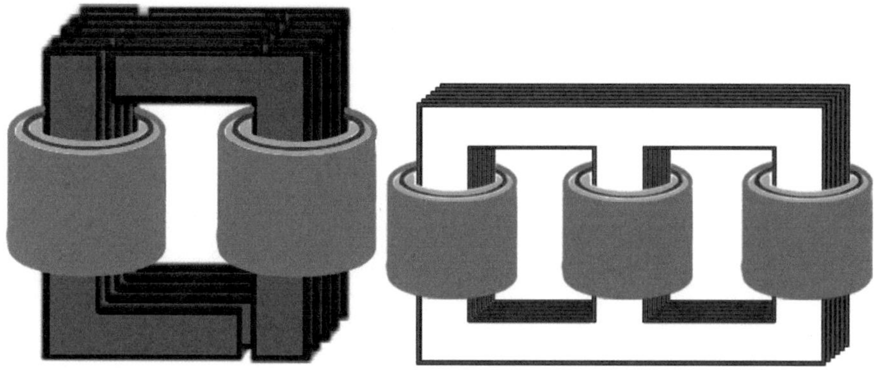


**Fig. 1.7** Single-phase and three-phase core transformer

### 1.3.3 Based on Construction

***Core-Type Transformer***: Basic windings construction of the core-type transformer encircles the core as shown in Fig. 1.7a and b. The limbs are coiled around the primary and secondary windings. As a result, it produces a single magnetic circuit. The high-voltage (HV) winding is looped outside, encircling the LV winding and removing it from the core, while the low-voltage (LV) winding is looped inside, close to the core. This winding design is justified by the fact that less insulation material is needed. Two windings can be tightly coupled together, with each limb carrying half of the primary and half of the secondary, to have less leakage reactance.

***Shell-Type Transformer***: In a trendy shell-type transformer, the core of the transformer encircles by winding as shown in Fig. 1.8a and b. The primary and secondary windings are sited together alternatively, giving rise to a sandwich-type winding with high-voltage windings sandwiched between low-voltage windings. In those conditions within the central limb, full flux passes and the remaining limb carries half flux. This creates a double magnetic circuit.

The assessment between core- and shell-type transformers is represented in Table 1.3.

### 1.3.4 Based on the Voltage Ratio

- ***Step-Up Transformer***: This type of transformer increases the voltage, with a corresponding decrease in current. They have a larger number of turns on the secondary side, i.e., $(T_2/T_1) > 1$, where $T_1$ and $T_2$ are primary and secondary turns

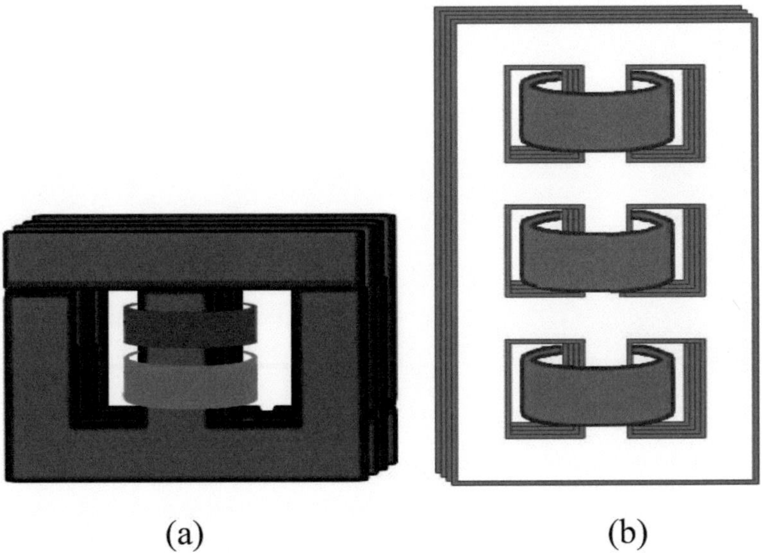

Fig. 1.8 Single-phase and three-phase shell-type transformer

**Table 1.3** Comparison between core-type and shell-type transformers

| Feature\type | Core type | Shell type |
|---|---|---|
| Basic construction | Simple, as the core is encircled by the windings | Complex, as the core encircles the maximum part of the windings |
| Number of the magnetic circuit | Single | Double |
| Number of limbs (for single-phase type) | Two | Three |
| Type of coil preferred | Cylindrical | Multilayer disk or sandwich-type |
| Leakage reactance and possibility of reduction | Comparatively high and cannot be reduced | Comparatively low and can be reduced |
| Maintenance and repairs | Easier to maintain and repair as winding can be removed or dismasted for repairs | Difficult to maintain and repair as for dismasting, a large number of laminations are required to be removed |
| Cooling and heat dissipation | Effective, due to the distribution of windings | Difficult, as mainly the core can be effectively cooled rather than the windings where more heat is produced |
| Electromagnetic forces experienced under fault or short-circuit conditions | Large | Small |
| Applications | Suited for high-voltage applications | Suited for low- to medium-voltage applications |

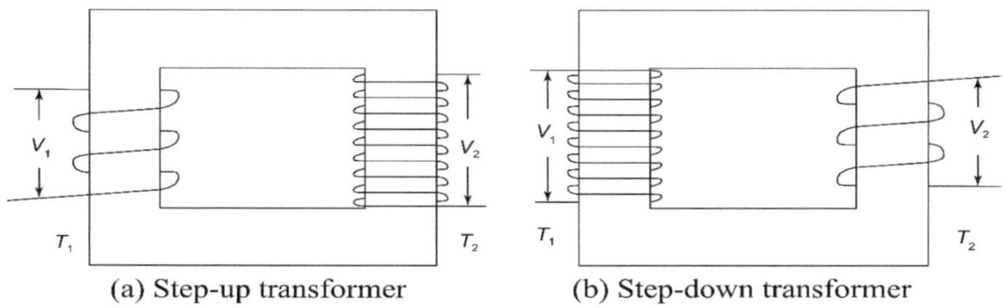


**Fig. 1.9** Step-up and step-down transformers

of the transformer, respectively. A view of the step-up transformer is revealed in Fig. 1.9a.

- ***Step-Down Transformer***: This type of transformer decreases the voltage, with the corresponding increase in current. They have fewer secondary side turns, i.e., $T_2/T_1 < 1$, where $T_1$ and $T_2$ are the primary and secondary turns, correspondingly. A view of the step-down transformer is particularized in Fig. 1.9b.

Some of the different transformer assemblies are illustrated in Fig. 1.10.

## 1.4 Different Parts of the Transformer

Transformer design aims to obtain the measurements of all active and inactive parts to supply these data to the customer. Some of those important parts, like winding, core, transformer tank, tap changer device, bushings, coolers, etc., are shown in Fig. 1.11. Each part has a precise task and accountability. For the manufacturing and assembling of these objects, different engineering techniques are applied to model the complete transformer [2]. Each part has a specific task and responsibility.

### *1.4.1 Winding*

The winding is called the heart of the transformer, which is made up of copper and aluminum. It is the costliest element of a transformer. Transformer windings are designed to meet three primary requirements, viz. mechanical, thermal, and electrical. They are cylindrical in shape and are assembled concentrically. To make winding, paper-insulated high conductivity and soft-drawn copper conductor are being used in compliance with national and international standards. In accordance with the application and rating of the transformer, the winding type is decided by the designer. Cylindrical windings, helical windings, crossover windings, continuous disk winding, interleaved winding, intra-shield winding, foil winding, etc., are the

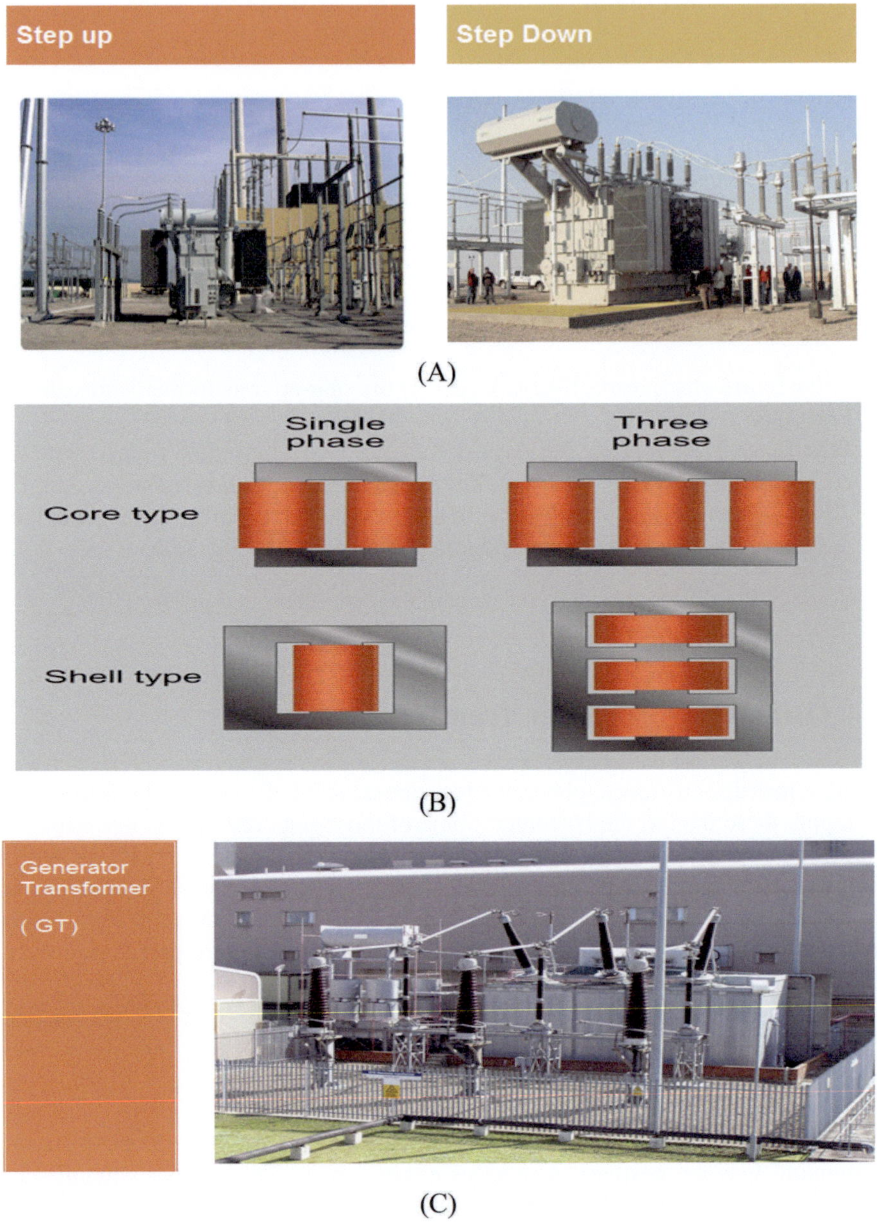


**Fig. 1.10** Different types of transformers. **a** Step-up and step-down transformer, **b** core- and shell-type transformer, **c** generator transformer, **d** power and distribution transformer, **e** single- and three-phase transformer, **f** furnace- and dry-type transformer, **g** traction and rectifier duty transformer, **h** testing transformer

(D)

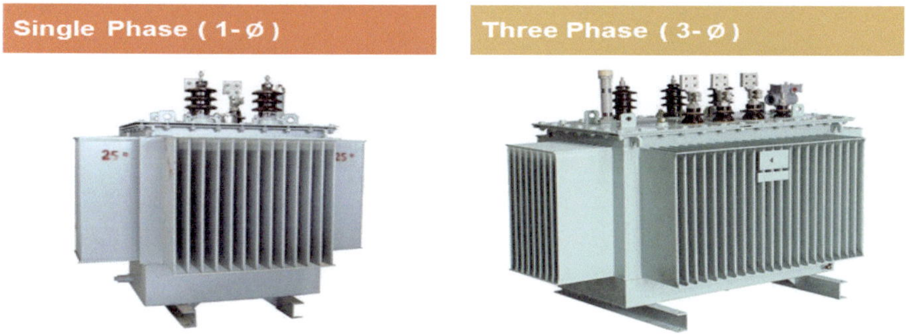


(E)

(F)

**Fig. 1.10** (continued)

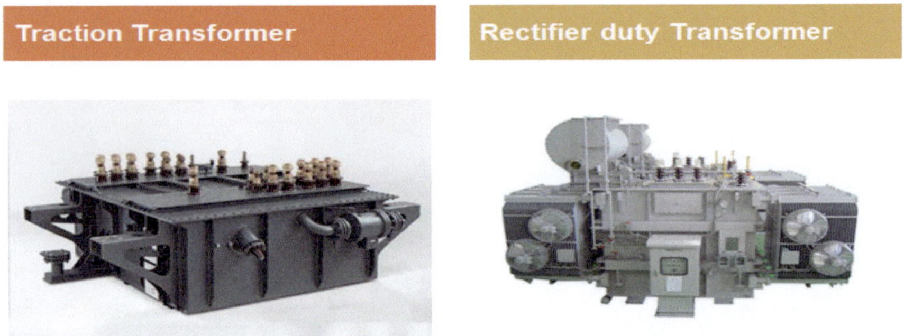


(G)

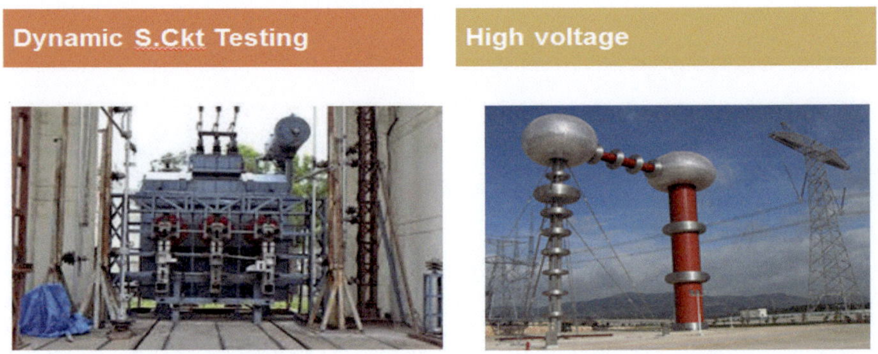


(H)

**Fig. 1.10** (continued)

different types of winding construction. For example, the interleaved or intra-shield windings are used for EHV or UHV class transformers, while helical windings are used for high-current transformers. Figure 1.12 demonstrates the winding and core assembly in actual form.

### *1.4.2 Core*

The core forms the magnetic circuit of the transformer; it is also the costliest component of the transformer after the winding. It is the heaviest part of the transformer. The core is constructed with cold-rolled grain-oriented sheet steel (CRGO). CRGO plates are generally used thickness of 0.23 or 0.27 mm, which are stacked one over the other as per standard design practices. To reduce losses and noise, the core is tightened with binding material like fiber tap, glass resin tape, etc.

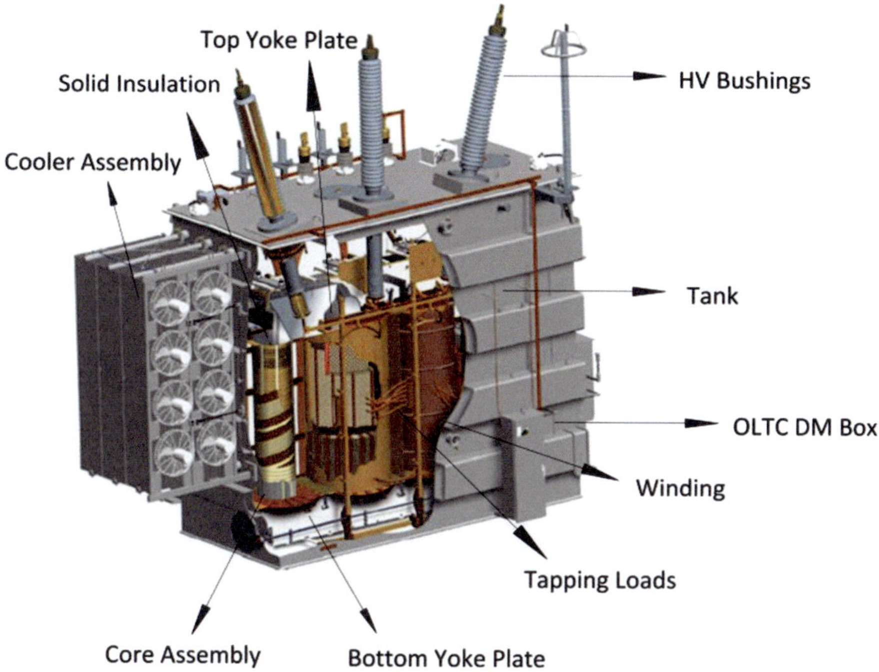


**Fig. 1.11** Composite view of different parts of transformer [2]

**Fig. 1.12** Transformer winding and core assembly

## *1.4.3 Tank and Radiator*

Tank is considering either housing or enclose for the transformer active part which is made from mild steel. The structure of the tank must be robust to protect against the severe mechanical stresses during operation of transformer in the power system network. All the accessories including coolers and radiator are mounted on the tank. It is also one of the heaviest parts of the transformer. For the best quality of transformer tank, it passes through key tests like pressure test, vacuum test, radiography

**Fig. 1.13** Transformer tank and radiators

test, etc. Furthermore, those radiators are the component of the transformer that is responsible to provide cooling to the transformer. These are being made from the CRCS steel plates which are spited up into a numbers of panels as per the designer of the transformer. It also passes a rigorous leak test to avoid leaking while functioning with the transformer. Figure 1.13 demonstrates the actual view of the tank and radiator.

### *1.4.4 Bushings*

Bushings are the link between transformer's active part and the transmission line that means electricity either enters or comes out from the transformer only through bushings. Without a bushing, a transformer connection with the electrical network is not possible. It is made up of material like porcelain, silicon rubber etc. The widely used types of bushings are condenser type, solid porcelain type, Resin Impregnated Synthetic (RIS), Resin Impregnated Paper (RIP), epoxy casting type, etc. Figure 1.14 demonstrates the different types of bushing used in a transformer.

### *1.4.5 Insulation Component*

The insulation coordination is a prime important engineering for transformers. Figure 1.15 shows the required different types of insulation in the transformer. Without Insulation components, the active parts construction is not possible to cater voltage and current stress. All key performance parameters of transformers like % impedance, basic insulation level, dynamic short-circuit withstand capability are greatly influenced by this insulation system of the transformer. Pressboards, compressed wood, crepe and craft papers, Nomex Paper, fiberglass, etc., are the

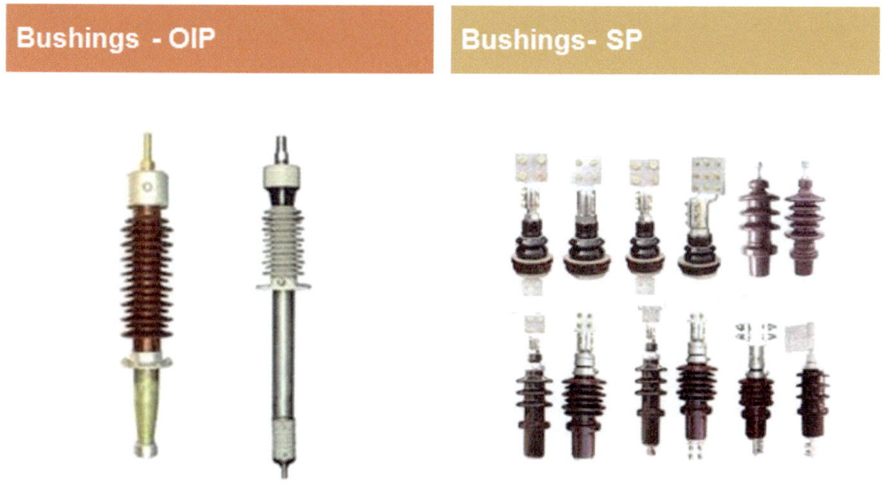


**Fig. 1.14** Transformer bushings (condenser and solid porcelain type)

various insulations being used in transformers or laminated wood. It provides the mechanical as well as electrical strength to the transformer against the radial-axial forces and high-voltage strokes.

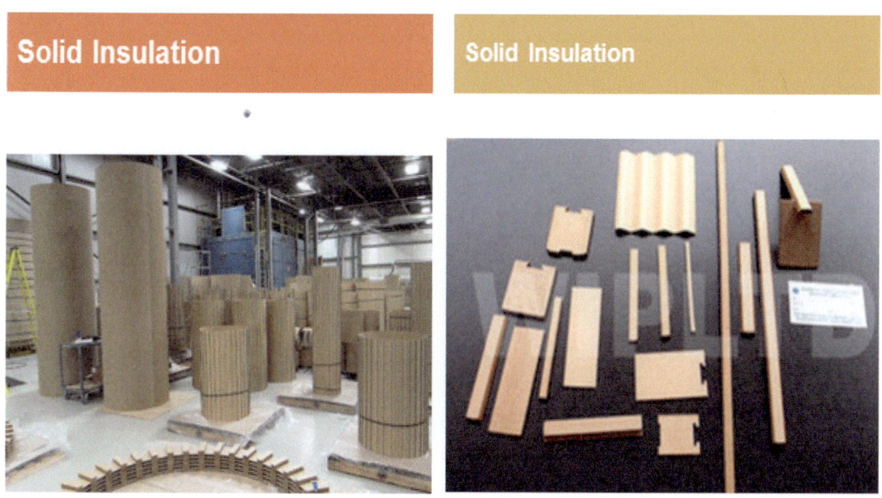


**Fig. 1.15** Insulation components for transformer

### *1.4.6 Tap Changing Device*

The voltage variation is a frequent phenomenon in the electricity grid network due to sudden rise and fall of load and which results in an increase or decrease in the transmission voltage. For the proper functioning of electrical equipment, constant voltage is of utmost importance, which is made possible by using the tap-changing device in the transformer. There are mainly two types of tap change widely used for the transformer, on-load and off-load tap changers. Inner and outer views of OLTC are revealed in Fig. 1.16.

## 1.5 Transformer Materials

The transformer is being manufactured with hundreds of raw materials like copper, CRGO, oil, insulation components, and many more. These components are also manufactured as per national and international standards with a strong quality check to fulfill the customer's requirement. A few of the key materials are described in detail herewith.

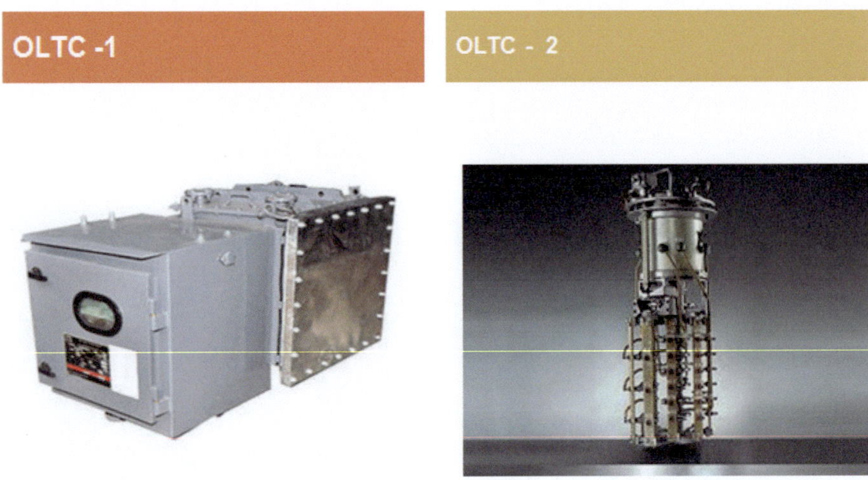


**Fig. 1.16** Tap changer for transformer (OLTC and OCTC)

**Fig. 1.17** Copper and lamination

### *1.5.1 Copper*

Copper is a widely used, highly conducting, and low-resistive material which is well suitable as per engineering norms for the transformer design and manufacturing. However, aluminum is also being used for a certain class of transformer. Copper or aluminum is used for the construction of winding and connection. Mainly, rectangular and round conductor shape is used across the industry for the ease of manufacturing. For more than a decade, individual rectangular conductors have been replaced by continuously transposed conductor (CTC), which is the best solution for achieving the mechanical strength of winding and reduction of eddy current/stray losses. This copper conductor comes with insulation covering as advised by the transformer manufacturer. Copper alone covers major portion of the raw material cost of transformer. Different copper and laminations (like, CRGO material) are shown in Fig. 1.17.

### *1.5.2 Cold-Rolled Grain-Oriented (CRGO) Steel*

CRGO is also one of the key costly raw materials for transformer manufacturing which comes in rolled sheet steel form. The thickness of CRGO varies from 0.23 to 0.3 mm, and around 3% silicon content is added during the grain orientation process to increase of electrical resistivity and minimization of eddy current losses, which finally increases the overall efficiency of the transformer. This material is available from specific countries, so it is considered one of the influencing materials for the manufacturing of transformers.

### 1.5.3 Solid Insulating Material

The life of the transformer depends on the condition of the insulation used for it. Insulation components are used for winding construction, core construction and assembling of core and winding together. The voltage-stress model and thermal model of the transformer are mainly referred to the insulation system employed. The insulating materials like pre-compress pressboard (PCPB), compressed wooden, Kraft and Crepe paper, etc., are as shown in Fig. 1.18. These materials are mostly considered cellulose based material because made from pulp and wood so it is considered hygroscopic in nature. That means it absorbs moisture more frequently once it comes in contact with that atmosphere. So, these components must be moisture-free (dry) before putting the transformer in service. Insulation coordination is one of the main topics which must be taken care of prudently for the life expectancy of the transformer.

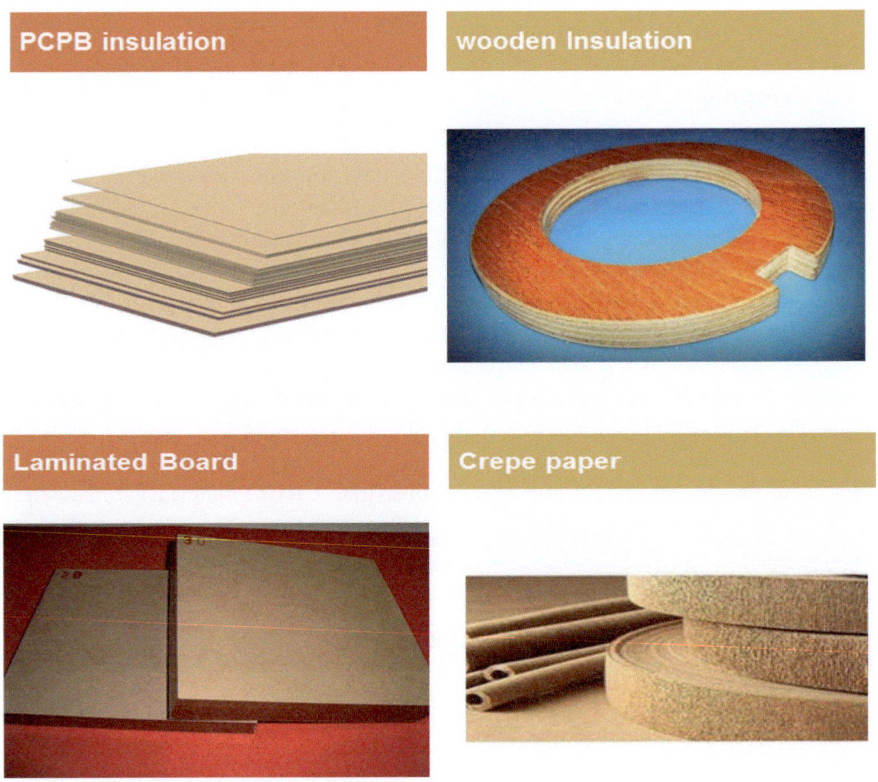


**Fig. 1.18** Cellulose base insulating material

**Fig. 1.19** Dielectric oil for power transformer

### 1.5.4 *Insulating Oil*

Insulating oil is mainly a petroleum product that is tuned up to achieve electrical insulation characteristic. It is known as transformer oil (shown in Fig. 1.19). This oil is used in the transformer for mainly two purposes. One is as insulation and the other as a coolant. It is a liquid insulating material used in a transformer that has high electrical insulating and heat, and mass transforming properties. On the principle of thermodynamics, oil rotates inside the tank and from corner to corner of the active component and extracts the heat and that heat is cooled down through the route of radiators fixed on the tank of the transformer. The insulation resistance of a transformer greatly depends on the properly of transformer oil. The key characteristics of transformer oils are breakdown voltage (BDV), resistivity, density, viscosity, etc. Nowadays, plant-based ester oil and synthetic-based ester oils are being used as a substitute of conventional petroleum-based transformer oil in context of environmental and safety concern.

### 1.5.5 *Mild Steel*

The important components of the transformer, like tank, core clamp, bushing housings, etc., are made from mild steel raw material. So, the grade of steel is an important factor for the strength and efficiency of these components. For example, if the steel grade is lower, the tank might rupture during the dynamic pressure build in the tank due to the short-circuit phenomenon. On the other side, stray loss can increase dramatically due to more accumulation of flux on the tank. In the context of the core clamp and bushing housing temperature increased due to stray and leakage flux

**Fig. 1.20** Mild steel for transformer tank fabrication

deposited on the steel body. The transformer core-winding assembly is to be housed in the tank and also to be filled with oil. The tank is made up of mild steel to provide protection to the CCA against the pressure and vacuum generation due to temperature rise while working. During the course of fault and failure, it plays a vital role in handling the developed excessive pressure and prevents the chance fire in yard. View of mild steel is expressed in Fig. 1.20.

The design of the transformer is happening as per the customer's technical specification and using available materials economically in order to achieve low cost and optimized performance parameters. The transformer design is worked out using various methods based on accumulated experience realized in different formulas, equations, tables, charts, and data.

## 1.6 Transformer Design Philosophy

Designing a transformer is a highly intricate task. When creating the core and winding, several crucial parameters come into play, including impedance (%Z), efficiency, flux density, and current density, as well as no-load and load losses, all of which are influenced by the construction method and the materials chosen for the core and winding. The initial step in designing the core involves selecting the flux density and the material grade. Various core materials are available on the market, such as CRGO and Amorphous. CRGO comes in multiple grades, including M4, M5, M6, ZDkh, Zh90, MOH, and Z0h. To reduce losses in core materials, the silicon steel content should be between 3 and 4%. Laminations are used in individual core plates to minimize eddy current losses. Conversely, designers use different types of windings tailored to various kV classes and current ratings. These can include helical winding, crossover coils, layer winding, continuous disk winding, interleaved disk winding, and counter shield winding, suited for current ratings of 6000 A, 1200 A, and 40 A, respectively, in the design of transformers tailored to customer requirements, operating voltage classes, and kVA ratings. Prior to transformer construction, the manufacturer prepares a computer-based transformer design and calculates the design parameters. In the past, manufacturers typically used Excel sheets to design transformers. However, nowadays, various software programs are available for this purpose. After all parameters are calculated, the transformer design moves into production, where the transformer is manufactured and subjected to various tests to confirm its design. Research in transformer engineering and design optimization is expanding significantly. Numerous studies are currently being conducted in this domain. Transformers go through a detailed manufacturing process, followed by multiple tests to ensure that their design and performance are validated. The field of transformer engineering and design optimization continues to grow, with many investigations taking place. Designing a transformer remains a complicated task due to the necessity of balancing various conflicting needs, such as electrical ratings, efficiency, temperature rise, cost-effectiveness, and responsible use of natural resources.

A range of methodologies, including analytical methods and software-based strategies, are utilized in the design of transformers. Manufacturers and designers concentrate on optimization techniques aimed at satisfying customer requirements by either maximizing or minimizing certain objective functions while complying with various constraints. These objective functions can involve minimizing costs, losses, weight, material expenses, and production costs, or maximizing the transformer's rated power. The design of transformers goes beyond merely meeting customer needs; it also involves adhering to national and international standards such as IS, IEC, BS, and IEEE. This book delivers an in-depth examination of transformer design methodologies and optimization strategies. It aims to achieve an optimal transformer design according to specified requirements by efficiently using available materials to lower costs, decrease size, and improve operational efficiency. Various optimization techniques have been investigated, and a MATLAB-based optimization tool has been utilized to identify the most effective design parameters. The suggested approach and the efficacy of the MATLAB-based tool have been confirmed through the development of a prototype transformer, showcasing its practical usefulness.

## 1.7 Key Performance Parameters for Transformer Design and Development

The performance parameters encompass short-circuit impedance, no-load loss, load loss, stray losses, noise, efficiency, and more. Transformer design aspects, such as cooling systems, insulation coordination, and tank construction, are dependent upon certain parameters. Employing optimization techniques is the most effective approach for transformer design and development. For transformer design optimization (TDO), several methods are widely used.

- Arithmetic methods include producing various tools or software that simulate transformer design.
- Probability-based model which includes artificial intelligence approaches such as genetic algorithms (GA), artificial neural networks (ANN), swarm intelligence, and ant and bee algorithms. Certain methods are especially popular to use so as to optimize the transformer design.
- Although this approach still exists in the manufacturing industry for computing various parameters, they use equivalent circuit.
- Utilizing past experiences and relevant databases is also a method employed by some manufacturers and designers, who establish certain parameters to derive performance characteristics from their historical data.

Over the last few decades, several researchers have focused on specific characteristics. Some notable examples include the following.

### *1.7.1 No-Load Losses*

The fixed losses of a transformer remain unchanged regardless of its loading conditions. No-load loss, also known as core or iron loss, comprises two components: hysteresis and eddy current loss [2]. Additionally, there is a stray loss component, which cannot be attributed to either eddy or hysteresis phenomena. The evaluation of eddy current losses in tanks [3] and stray losses in windings, core, tank, and clamps is conducted using finite element method (FEM) [4]. One approach discussed in [5] involves the development of correction factors for 2D FEM calculations in a 3D context. A novel software application for optimizing transformer cores to assess flux and iron losses is detailed in [6]. In [7], artificial neural networks (ANNs) are utilized to predict iron losses in transformer cores. The ANN-GA scheme is proposed for minimizing no-load losses in [8]. Meanwhile, [9] illustrates the reduction of no-load losses through the application of FEM. The use of FEM has been extensively documented concerning no-load losses. The losses associated with the flitch plate are effectively showcased in [10] through both 2D and 3D FEM simulations. Research focused on local flux concentration and distribution in transformer cores, particularly concerning joint design and its effectiveness in addressing losses and noise, is covered in [11]. A combination of ANNs, decision trees (DTs), and genetic algorithms (GAs) is presented in [12] to minimize core losses. The creation of an enhanced equivalent circuit for precise loss measurement is proposed in [13]. An experimental investigation into various samples to evaluate different stacking patterns, directional flux lines, and slitting angles is significantly relevant to reducing core losses in power transformers [14].

### 1.7.2 Load Losses and % Impedance

Losses of this nature arise from the load current that passes through the transformer [2]. Copper losses or winding losses are also called the losses that can occur in these. They can alternatively be defined as the sum of two losses I2R losses and stray losses, where I represents the current flowing through the winding and R represents the resistance of the winding material of copper or aluminum. FEM is applied to determine a current "I" for a distribution transformer [15]. When the transformer is under load conditions, stray losses take place. From the point of view of the field of transformer engineering, these stray losses are a serious matter. Kulkarni and Khaparde [16] reviewed about 50 papers on stray loss investigation and estimates or mitigation. Stray losses occurring at clamping plates, flange bolt areas and bushings or throats have been reduced by using Finite Element Method (FEM) [17–19]. The impact of eddy currents resulting from windings and high-current leads in large transformers is examined using the FEM method [20]. The 3D reluctance method is suggested for predicting load losses in transformers [21]. 2D FEM continues to be an effective tool for evaluating leakage fields and short-circuit impedance, as highlighted by recent advancements in [22]. A formula-based approach for calculating transformer leakage fields is proposed in [23], detailing the calculations of both self and mutual impedances.

### 1.7.3 Cooling Process

The cooling process of a transformer is a crucial factor as it directly impacts the transformer's longevity. Transformer engineers take this responsibility to design an effective cooling system. The advent of thermal modeling has converted transformer engineering from earlier methods and conventional analytical calculation processes. In [24] a discussion has been done regarding hotspot temperatures in transformer cores, and they state something in terms of a "2D FEM thermal model." Reference [25] outlines the thermal dynamics with regard to the circulation of the cooling medium required for heat and mass transfer in order to improve thermal efficiency. The methodology for calculating the rise in top oil temperature and hotspot temperature is examined in [26]. The mechanism of thermal loading and corresponding guidelines are provided in [27], based on experimental results, factory tests, and field data measurements. A mathematical model focusing on hotspot temperatures in windings in relation to load current is introduced in [28]. A genetic algorithm (GA)-based thermal model for power transformers has been developed and is discussed in [29]. The hydrodynamic characteristics of oil enable oil circulation or flow in ONAN transformers, as noted in [30]. A proposal for the correlation between load and ambient temperatures, taking seasonal variations into account for substation transformers, is presented in [31]. The approach for assessing cyclic loading in power transformers is detailed in [32].

## 1.8 Transformer Design Optimization (TDO)

The optimization of transformer design has become essential for manufacturers to reduce both manufacturing and material costs while meeting customer specifications. For design engineers, optimization presents a significant challenge. When an area for optimization is identified, the first step is to gather comprehensive information about the issue at hand. Following this, consideration must be given to objective functions that either aim to maximize or minimize under certain constraints. For optimization of transformer design optimization, various algorithms need to be implemented to create the ideal cost that would be optimized by the algorithms and align with mechanical, electrical, and thermal conditions. The main objective is to design the dimensions of all transformer elements based on the customer specifications to the manufacturer [33]. Optimizing transformer design is a complex endeavor, requiring researchers to ensure compliance with necessary specifications while keeping production costs low. An optimal transformer design guarantees the manufacturer a product that delivers high performance at a comparatively low expense. The design can be optimized by focusing on reducing manufacturing costs mass, or losses, or by enhancing transformer apparent power, total ownership cost, transformer active part cost, or active part mass, all while adhering to specified requirements.

### *1.8.1 Manufacturing Cost Minimization*

For optimizing transformer design, the primary objective is to lower manufacturing expenses. By using first-class design software, this can be easily carried out. It was done on a computer in 1955 and completed as an initial transformer design. Research to design transformers using computational methods was then started. Early attempts relied on adapted analytical solutions, enabling the optimization of their construction and leveraging advances in the properties of magnetic and electric materials, along with trial and error methods for design validation. For instance, costs can be reduced by altering the winding material or modifying other physical parameters. The application of high-temperature superconductor as winding material is achieved in design optimization for a 3.5 MVA power transformer in [34]. Many transformer manufacturers still apply conventional analytical design techniques out to this day. It is a process that takes time to reach the optimal design. The drawback with this is that many parameters need to be simultaneously manual reviewed. The literature very rarely addresses the overall minimization of manufacturing costs; instead, it often aims to minimize the component costs [35].

### 1.8.2 Operating Cost Minimization

Alongside the manufacturing expenses, the total ownership cost (TOC) is another factor for transformer design optimization (TDO), which includes purchase costs and energy losses. The idea of minimizing TOC for distribution transformers is discussed in sources [36]. To achieve the most cost-effective transformer, TOC should be given serious consideration [37]. The influence of the environment on transformers and the impact of transformers on the environment must also be taken into account [38].

### 1.8.3 AI and ANN Techniques for TDO

Various techniques have been utilized by researchers to enhance the design of transformers for different transformer types. Geometric programming (GP) serves as an analytical method for addressing a specific format of optimization problems. GP is recognized as one of the earliest optimization applications. The design challenge has been represented as a geometric program and resolved using the related analytical approach [39]. The problem is to find a minimum capitalized cost of power transformer geometrically and electrically optimally adjusted three-winding, three-phase core form powcr transformcr. No-load losses, load losses, core mass, and short-circuit impedance are constrained. Moreover, in [40], the authors have seen that neural network techniques are also employed in the design optimization of the transformer. The most straightforward approach, referred to as the iterative or exhaustive search method, involves thoroughly examining all possible design variable values to discover the solution that yields the optimal value for the objective function. The restriction of total losses, no-load losses, and the short-circuit impedance is considered along with the design variables of the height and width of the core, the cross-sectional areas of HV and LV winding conductors, and flux density [33]. Similarly, a methodology is applied to the minimization of the active part of a three-phase core-type distribution transformer with K-factor, flux density, and current density as design variables subject to constraints that include no-load losses, total losses, and percent impedance put on a three-phase core-type distribution transformer as described in [41]. The problem of optimizing a single-phase shell-type transformer with multiple objectives such as ownership cost, mass, total losses, material cost, efficiency, and impedance for a 25 kVA transformer is solved in [42]. Although, this multi-objective approach can be quite time consuming because of many parameters to be governed simultaneously for the design optimization. A method based on the decision tree in [43] was implemented to get a quicker result. It involves several decisions depending on no-load condition, no-load losses, load condition, load losses, and the rated power. Each design variable is narrowed down to the ranges and ten random values are selected and tested from each interval for optimality. A number of winding turns is determined in [44] by using an ant colony algorithm.

## 1.9 Conclusion

Transformers are indispensable units of power systems as it is the main switch between generation transmission and distribution networks. The whole stability and economic feasibility of an electricity supply system will be determined by how directly they will be designing and actually performing their systems. It is important not only to understand the parameters of key performance such as efficiency, percentage impedance, no-load losses, and load losses in order to have effective transformer operation to make intelligent decisions while on designing and procurement of the transformer. Transformer core, windings, insulation system, and cooling mechanisms are important structural constituent of a transformer, which govern its reliability and life. Selection of material for the core steel and the insulating oil has a great influence on losses and thermal performance. Particularly, artificial neural networks (ANNs) have been effective at solving problems of loss estimation, predictive modeling, and design optimization. In addition, it is possible to achieve these dual goals of minimizing manufacturing and operating costs with greater ease through the deployment of simulation-based design and AI-based optimization algorithms. Finally, transformer engineering develops and continues to respond to traditional practices coupled with new innovations. To advance next-generation performance-optimized transformers suitable for the complexities of modern energy systems, the power industry has the opportunity and flexibility to adopt such a holistic approach to bring to the market a combination of best performance, lowest cost, and lowest maintenance transformers which incorporate material science, cost analysis, and apply learning methodologies using AI.

## References

1. IEEE draft standard terminology for power and distribution transformers. IEEE PC57.12.80/D6, 2010, pp 1–55
2. Ushakov VY, Mytnikov AV, Rakhmonov IU (2023) Characteristics of the main elements of electric power systems. In: High-voltage equipment of power systems. power systems. Springer, Cham. https://doi.org/10.1007/978-3-031-38252-9_2
3. Williams SB, Abetti PA, Magnusson EF (1956) Application of digital computers to transformer design [includes discussion]. Trans Am Inst Electr Eng Part III Power Appar Syst 75(3):728–735. https://doi.org/10.1109/AIEEPAS.1956.4499359
4. Guerin C, Tanneau G, Meunier G (1993) 3D eddy current losses calculation in transformer tanks using the finite element method. IEEE Trans Magn 29(2):1419–1422. https://doi.org/10.1109/20.250669
5. Ngnegueu T, Meunier G, Sabonnadiere JC, Mailhot M, Arthaud JP (1990) 2-D finite element calculation of shell losses in power transformers, taking the third dimension into account by means of a correction factor. IEEE Trans Magn 26(2):925–928. https://doi.org/10.1109/20.106469
6. Yu CH, Basak A (1993) Optimum design of transformer cores by analysing flux and iron loss with the aid of a novel software. IEEE Trans Magn 29(2):1446–1449. https://doi.org/10.1109/20.250675

7. Kollias SD, Georgilakis PS, Hatziargyriou ND, Doulamis ND, Doulamis AD (1998) Prediction of iron losses of wound core distribution transformers based on artificial neural networks. Neurocomputing 23(1–3):15–29
8. Guangzheng N et al (1992) FEM analysis of 3-D transformer leakage field and eddy current loss in the windings. IEEE Trans Magn 28(2):1382–1385. https://doi.org/10.1109/20.123950
9. Souaris A, Paparigas DG, Tsivgouli AJ, Tsili MA, Georgilakis PS, Kladas AG (2006) Finite element methodology for the accurate prediction and minimization of no load loss in power transformers. Gen Secr Res Technol Greece under Grant 00BE457, pp 510–516
10. Koppikar DA, Kulkarni SV, Srinivas PN, Khaparde SA, Jain R (1999) Evaluation of flitch plate losses in power transformers. IEEE Trans Power Deliv 14(3):996–1001. https://doi.org/10.1109/61.772345
11. Swift G (1975) Excitation current and power loss characteristics for mitered joint power transformer cores. IEEE Trans Magn 11(1):61–64. https://doi.org/10.1109/TMAG.1975.1058541
12. Georgilakis P, Hatziargyriou N, Paparigas D (1999) AI helps reduce transformer iron losses. IEEE Comput Appl Power 12(4):41–46. https://doi.org/10.1109/67.795137
13. Elleuch M, Poloujadoff M (2003) Analytical model of iron losses in power transformers. IEEE Trans Magn 39(2):973–980. https://doi.org/10.1109/TMAG.2003.808591
14. Findlay R, Belmans R, Mayo D (1990) Influence of the stacking method on the iron losses in power transformer cores. IEEE Trans Magn 26(5):1990–1992. https://doi.org/10.1109/20.104594
15. Pern J-F (1995) Calculating the current distribution in power transformer windings using finite element analysis with circuit constraints. IEE Proc Sci Meas Technol 142(3):231–236(5) [Online]. Available at: https://doi.org/10.1049/ip-smt_19951596
16. Kulkarni SV, Khaparde SA (2000) Stray loss evaluation in power transformers—a review. In: 2000 IEEE power engineering society winter meeting. Conference proceedings (Cat. No. 00CH37077), vol 3, pp 2269–2274. https://doi.org/10.1109/PESW.2000.847708
17. Olivares-Galvan JC, Magdaleno-Adame S, Escarela-Perez R, Ocon-Valdez R, Georgilakis PS, Loizos G (2014) Reduction of stray losses in flange-bolt regions of large power transformer tanks. IEEE Trans Ind Electron 61(8):4455–4463. https://doi.org/10.1109/TIE.2013.2279373
18. Olivares-Galvan JC, Campero-Littlewood E, Hernandez-Avila JL, Escarela-Perez R, Adame SM, Theocharis AD (2012) Evaluation of stray losses in throats of distribution transformers using finite element simulation. In: 2012 VI Andean Region international conference, pp 7–10. https://doi.org/10.1109/Andescon.2012.12
19. Tsivgouli AJ, Tsili MA, Kladas AG, Georgilakis PS, Souflaris AT, Skarlatini AD (2007) Geometry optimization of electric shielding in power transformers based on finite element method. J Mater Process Technol 181(1–3):159–164. https://doi.org/10.1016/J.JMATPROTEC.2006.03.049
20. Renyuan T, Yan L, Feng L, Lijian T (1996) Resultant magnetic fields due to both windings and heavy current leads in large power transformers. IEEE Trans Magn 32(3):1641–1644. https://doi.org/10.1109/20.497569
21. Turowski J (2000) Fast 3-dimensional interactive computation of stray field and losses in asymmetric transformers. IEE Proc Gener Transm Distrib 147(4):197–201(4) [Online]. Available at: https://doi.org/10.1049/ip-gtd_20000435
22. Cui X, Zhang G (2000) Automatic design of impedance in shell-form power transformer. IEEE Trans Magn 36(4):1822–1825. https://doi.org/10.1109/20.877799
23. Conlon M (1989) Calculation of self and mutual impedances between sections of transformer windings. IEE Proc C (Gener Transm Distrib) 136(5): 308–314(6) [Online]. Available at: https://doi.org/10.1049/ip-c.1989.0041
24. teNyenhuis EG, Girgis RS, Mechler GF, Zhou G (2002) Calculation of core hot-spot temperature in power and distribution transformers. IEEE Trans Power Deliv 17(4):991–995. https://doi.org/10.1109/TPWRD.2002.803703
25. Mufuta JM, Van Den Bulck E (2000) Modelling of the mixed convection in the windings of a disc-type power transformer. Appl Therm Eng 20(5):417–437. https://doi.org/10.1016/S1359-4311(99)00034-4

26. Jardini JA, Brittes JLP, Magrini LC, Bini MA, Yasuoka J (2005) Power transformer temperature evaluation for overloading conditions. IEEE Trans Power Deliv 20(1):179–184. https://doi.org/10.1109/TPWRD.2004.835433
27. Tojo S (2006) Proposition of individual loading guide for power transformers. IEEE Trans Power Deliv 21(3):1383–1389. https://doi.org/10.1109/TPWRD.2006.876979
28. Nordman H, Rafsback N, Susa D (2003) Temperature responses to step changes in the load current of power transformers. IEEE Trans Power Deliv 18(4):1110–1117. https://doi.org/10.1109/TPWRD.2003.817516
29. Galdi V, Ippolito L, Piccolo A, Vaccaro A (2001) Parameter identification of power transformers thermal model via genetic algorithms. Electr Power Syst Res 60(2):107–113. https://doi.org/10.1016/S0378-7796(01)00173-0
30. Zhang J, Li X (2004) Coolant flow distribution and pressure loss in ONAN transformer windings. Part II: Optimization of design parameters. IEEE Trans Power Deliv 19(1):194–199. https://doi.org/10.1109/TPWRD.2003.820216
31. Li X, Mazur RW, Allen DR, Swatek DR (2005) Specifying transformer winter and summer peak-load limits. IEEE Trans Power Deliv 20(1):185–190. https://doi.org/10.1109/TPWRD.2004.837680
32. Nguyen TT (1995) Constrained optimisation procedure for evaluating cyclic loading of power transformers. IEE Proc Gener Transm Distrib 142(3):240–246(6) [Online]. Available at: https://doi.org/10.1049/ip-gtd_19951746
33. Baranowski JF, Hopkinson PJ (1992) An alternative evaluation of distribution transformers to achieve the lowest TOC. IEEE Trans Power Deliv 7(2):614–619. https://doi.org/10.1109/61.127057
34. Zuo X, Li H, Zhao B, Liu J, Lu B, Li C (2021) Optimization design of a 40-kJ HTSPPT module for inductive pulsed power supply. IEEE Trans Plasma Sci 49(8):2380–2386. https://doi.org/10.1109/TPS.2021.3097124
35. Pham TH, Salon SJ, Hoole SRH (1996) Shape optimization of windings for minimum losses. IEEE Trans Magn 32(5):4287–4289. https://doi.org/10.1109/20.538845
36. Georgilakis PS (2007) Decision support system for evaluating transformer investments in the industrial sector. J Mater Process Technol 181(1–3):307–312. https://doi.org/10.1016/J.JMATPROTEC.2006.03.040
37. Kommission E, Forschung G (2003) External costs: research results on socio-environmental damages due to electricity and transport. Off. for Off. Publ. of the Europ. Communities, Luxembourg
38. Vesterdal M, Svendsen GT (2004) How should greenhouse gas permits be allocated in the EU? Energy Policy 32(8):961–968. https://doi.org/10.1016/S0301-4215(03)00019-3
39. Coelho LDS, Mariani VC, Guerra FA, da Luz MVF, Leite JV (2014) Multiobjective optimization of transformer design using a chaotic evolutionary approach. IEEE Trans Magn 50(2):669–672. https://doi.org/10.1109/TMAG.2013.2285704
40. Amoiralis EI, Georgilakis PS, Kefalas TD, Tsili MA, Kladas AG (2007) Artificial intelligence combined with hybrid FEM-BE techniques for global transformer optimization. IEEE Trans Magn 43(4):1633–1636. https://doi.org/10.1109/TMAG.2006.892258
41. Mehta HD, Patel R (2014) Minimization of active part cost of a 100 kVA distribution transformer using exhaustive search method. Int J Innov Res Electr Electron Instrum Control Eng 2(2):1100–1104
42. Olivares-Galvan JC, Georgilakis PS, Escarela-Perez R, Campero-Littlewood E (2011) Optimal design of single-phase shell-type distribution transformers based on a multiple design method validated by measurements. Electr Eng 93(4):237–246. https://doi.org/10.1007/s00202-011-0211-9
43. Amoiralis EI, Georgilakis PS, Litsos E (2006) Methodology for the optimum design of power transformers using minimum number of input parameters [Online]. Available at: https://api.semanticscholar.org/CorpusID:16997082
44. Fouzaï M, Zouaghi T (2013) Ant Colony algorithm applied to power transformer optimization. In: 2013 international conference on electrical engineering and software applications, Mar 2013, pp 1–5. https://doi.org/10.1109/ICEESA.2013.6578410

# Chapter 2
# Design of the Core and Yoke Through Conventional Technique

The conventional design technique of transformer cores and yokes remains fundamental in ensuring optimal performance, efficiency, and cost-effectiveness of electrical transformers. This chapter explores the traditional methodology applied to the core and yoke design, considering both core-type and shell-type transformer configurations. Central to the design process is the selection of volts per turn, which directly influences the core cross-sectional area and the overall dimensions of the magnetic circuit. The flux density is carefully chosen within permissible limits to ensure minimal core losses while maintaining magnetic efficiency. Likewise, current density is selected based on thermal constraints and economic factors to achieve a balanced compromise between copper loss and conductor size. Emphasis is placed on the geometry of the core, with particular attention given to square and stepped core arrangements to optimize space and reduce leakage flux. The window and core proportions are determined to meet both electrical and mechanical design requirements, providing adequate space for windings while maintaining magnetic integrity. Lastly, the yoke design is addressed to ensure uniform flux distribution and structural stability, completing the magnetic circuit with minimal reluctance. The conventional design approach, though manual and iterative, offers valuable insights into transformer behavior and design optimization.

## 2.1 Introduction

The transformer core and yoke are the most critical components in the transformer, as they magnificently decide performance, efficiency, and reliability of a transformer. The transformer core and yoke design methodologies have evolved over time with researchers seeking different configuration, materials, and structural optimizations in order to enhance efficiency and decrease losses. It provides a summary on the conventional design approach of transformer cores and yokes, discusses the loss reduction



N. Chothani et al., *Advancement in Infrastructure and Design Technology of Power Transformer*, Studies in Infrastructure and Control,
https://doi.org/10.1007/978-981-96-9420-4_2

techniques, points out the technological advancements, and is further intended to provide an overview of future challenges and research direction.

The transformer core and yoke design follows well-accepted engineering principles allowing for minimizing of core losses, maximization of magnetic flux path, and strength to meet mechanical requirements. To enhance efficiency, one normally selects an appropriate core material, determines core geometry, and optimizes the yoke design. Ferromagnetic materials used in making transformer cores help in efficient conduction of magnetic flux. To minimize hysteresis and eddy current losses [1], silicon steel laminations are still the most widely used in material. It has been shown that amorphous metal cores also offer improved performance at lower core losses, which makes them suitable for high efficiency applications [2]. However, cost constraints and mechanical limitations limit the use of silicon steel laminations to the use of common large-scale transformers.

The influence of flux distribution and transformer efficiency is in turn determined by the core geometry. The most common configurations are core type, shell type, and toroidal [3]. Therefore, in a core-type transformer, the windings are wound over two vertical limbs which offer good cooling, mechanical strength, and ease of insulation. In particular, this design is appropriate for high-voltage applications as it is capable of holding large winding spaces. On the other hand, it encloses the windings of transformer within the core; it provides the better magnetic shielding and lower leakage flux. It is also convenient for the compact transformers where reduction of leakage reactance is of the primary importance [4]. Advancement of finite element method (FEM) simulations has been used to improve core configurations, thereby reducing flux leakage as well as increasing transformer efficiency [5].

The magnetic circuit must be completed and the yoke is critically important for this purpose to contribute to low reluctance and minimum flux leakage. Mechanical stability, plus core loss reduction, is enabled with a properly designed yoke. This approach produces conventional designs which focus on the way the yoke can be designed to minimize flux density variations and reduce energy losses [6]. Current study has also explored ways to minimize air gaps between core and yoke; however, this improves magnetic flux linkage while also helping to boost overall transformer efficiency [7]. Decreasing the lamination thickness of the yoke for the reasons that eddy current losses are decreased and it can improve transformer performance [8] is another optimization strategy.

Hysteresis and eddy current losses are the most important sources of loss in transformer; hence, one should take the measures to minimize them. The magnetic properties of the core material affect the hysteretic losses. High-grade grain-oriented silicon steel studies indicate the reduction of hysteresis loss by reduction in magnitude [9]. For reducing energy dissipation, promising results have been achieved in using controlled annealing processes in order to increase the grain orientation in the core material [10]. The thin laminations coated with insulating materials are used similarly to minimize the eddy current losses, which occur due to the induced current in the core and yoke [11]. The high-resistivity coatings (amorphous oxide layers) further help insulation properties and reduce eddy current formation [12]. Step lap core construction makes for another effective methodology as this helps in improving flux distribution and reducing localized saturation and also results in lower eddy current losses [13].

At present, the conventional transformer design techniques are very reliable; however, owing to continuous technological developments, new optimization strategies have been invented. Finite element analysis (FEA) has gained wide acceptance in transformer core design since engineer can predict or analyze electromagnetic field distribution and core losses with high precision [14].

Using artificial intelligence (AI) and machine learning, magnetic flux distribution is also optimized. Optimal flux paths are predicted using AI based models so as to reduce localized saturation and increase efficiency of the transformer [15]. They enable the engineer to automate core design processes diminishing the size of the transformer and use of material. Furthermore, core efficiency has also been improved by laser cutting and precision lamination stacking [16]. Lamination lacing by laser cutting and precision lamination shaping lower air gap and optimize magnetic flux linkage reducing whole losses of energy.

The conventional transformer design techniques are still used quite widely though with several drawbacks. Hysteresis, eddy current losses, and the corresponding magnetic energy loss from these materials are known to be reduced further by nanocrystalline and amorphous core materials [17], which might be explored in future studies. Finally, various aspects of smart monitoring systems integrated into transformer cores or yokes are another promising direction. Sensors and real-time monitoring systems can be embedded into the circuit, so that they facilitate detecting performance anomalies, which potentially leads to implementation of the predictive maintenance strategies and increasing of transformer lifespan [17]. Moreover, improved transformer efficiency and longevity could be achieved through better ventilation and cooling fluid designs. Even though there are higher scope for improvement of the efficiency with wider controlling ability in solid-state transformer (SST) [18–20], due to its protection for the higher range required to concentrate on the conventional transformer designing. Even protection of the conventional transformer also face many issues related to inrush [21, 22], CT saturation [23], higher resistance internal fault [24], overfluxing [25], and also malfunctioning of conventional percentage bias differential protection under certain abnormal conditions [26, 27].

Transformer cores and yokes continue to be an integral part of electrical engineering and that means balancing of magnetic efficiency, mechanical stability, and minimum energy loss. However, recently advances in the simulation of FEA and advancements in AI based optimizations, and precision manufacturing have started to redefine the way in which transformers are designed, using either traditional design techniques of laminated silicon steel, optimized core geometries and step lap constructions. To achieve high-performance transformers that meet the evolving needs of modern electrical systems, the material limitations, flux optimization, and cooling inefficiencies will be addressed in a forefront method.

## 2.2 Required Specifications of a Transformer Design

The transformer specifications are their rating and performance expectations. A designer needs to know the following important specifications with regard to the design of the transformer:

1. Rating/capacity/output—in kVA.
2. Voltage rating of HV and LV windings with or without tap changers and tapings.
3. Frequency—"*f*" in Hz.
4. Number of phases—one or three.
5. Star-star, star-delta, delta-delta, delta-star (with or without ground neutral) type of winding connection of three-phase transformers.
6. Continuous or short time rating.
7. Cooling—natural or forced
8. Type—core or shell, power or distribution.
9. It is efficiency, per unit impedance, location (i.e., indoor pole or platform mounting, temperature rise, etc.).

## 2.3 Design of Single-Phase and Three-Phase Transformers

The design of a transformer includes the determination of its main dimensions, namely the overall width and height, based on the geometry of windows, distance between limbs and diameter of the circumscribing circle for various types of core. Design also includes the design of yoke, core, and winding. Estimation of no-load current, iron loss, and copper loss facilitates the temperature rise calculations based on which cooling systems are designed. Calculation of resistances and leakage reactance is used for the determination of the efficiency and regulation.

Hence, the design of a transformer can be split into following four steps:

1. Determining the output equation relating to the main dimensions.
2. Determining overall dimensions.
3. Design of cooling apparatus.
4. Estimation of quantities such as no-load current, iron losses, resistance, leakage reactance, and so on.

### *2.3.1 Design of Single-Phase Transformer*

The kVA rating of a single-phase transformer is given as follows:

$$Q(\text{kVA}) = V_{\text{p}} I_{\text{p}} \times 10^{-3} \text{ (or) } V_{\text{S}} I_{\text{S}} \times 10^{-3}, \tag{2.1}$$

$$\cong E_{\text{p}} I_{\text{p}} \times 10^{-3} \left[\because \text{Terminal voltage, } V_{\text{p}} \cong \text{Induced voltage, } E_{\text{p}}\right] \tag{2.2a}$$

(or)

$$\cong E_S I_S \times 10^{-3}\ [\because V_S \cong E_S], \tag{2.2b}$$

where p → primary, $V$ and $E$ → voltage, s → secondary, and $I$ → current.

The voltage induced in a transformer winding is given as follows:

$$E_x = 4.44 f \varphi_m T_x, \tag{2.3}$$

where $f$—frequency (Hz), $\varphi_m$—Flux (Wb), $T$—number of turns, $x$—can be p or s, denoting primary or secondary.

### 2.3.2 Design of Single-Phase Core-Type Transformer

The cross-section of a single-phase core-type transformer is shown in Fig. 2.1. The arrangement of the winding is shown in Fig. 2.2.

The output of the transformer in kVA is related to its main dimensions. Hence, the output can be determined in terms of area of core/iron ($A_i$), area of window ($A_w$), and window space factor ($K_w$). Hence, the following steps are followed.

In case of single-phase transformers, there is one primary and secondary winding. Half of the primary and half of the secondary are placed on one limb and the other half of each of the windings is placed on the other limb. Each limb has half of the primary and secondary windings as shown in Fig. 2.10. Hence, the total copper area in the window is given as follows:

$$\begin{aligned} A_c &= \text{Total copper area of } [\text{primary winding} + \text{secondary winding}] \\ &= [\text{number of turns in primary} \times \text{area of primary conductor}] \\ &\quad + [\text{number of turns in secondary} \times \text{area of secondary conductor}] \\ &= T_p A_p + T_S A_S \end{aligned} \tag{2.4}$$

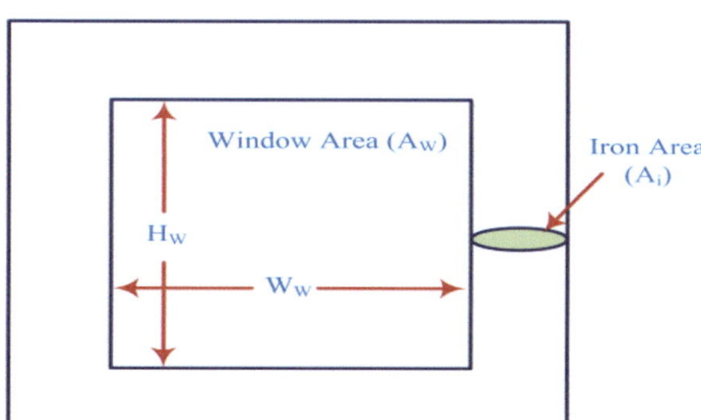


**Fig. 2.1** Cross-section of 1$\varphi$ core-type transformer

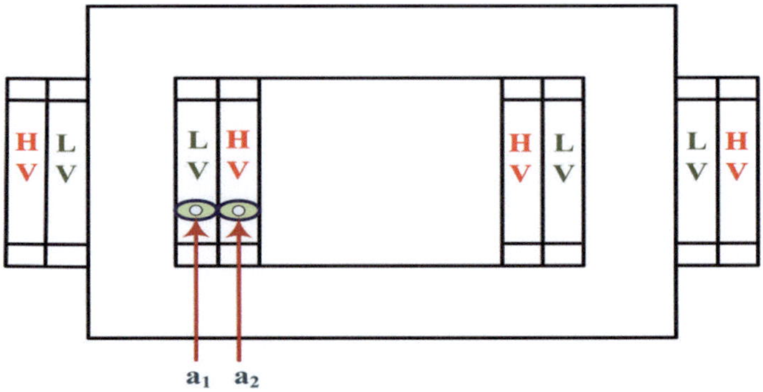


**Fig. 2.2** Winding arrangement of 1$\varphi$ core-type transformer

Assuming same current density, $\delta$ in both primary and secondary windings, we also get

$$\delta = \frac{I_p}{A_p} = \frac{I_s}{A_s} \tag{2.5}$$

$$\Rightarrow A_p = \frac{I_p}{\delta} \quad \text{and} \quad A_s = \frac{I_s}{\delta} \tag{2.6}$$

Substituting Eqs. (2.5) and (2.6) in Eq. (2.4), we get

$$A_c = T_p \frac{I_p}{\delta} + T_s \frac{I_s}{\delta} \tag{2.7}$$

Assuming,

$$T_p I_p = T_s I_s = \mathrm{AT} \tag{2.8}$$

where AT is the magnetic mmf.

Substituting Eq. (2.8) in Eq. (2.7), we get

$$A_c = \frac{\mathrm{AT}+\mathrm{AT}}{\delta} = \frac{2\mathrm{AT}}{\delta} \tag{2.9}$$

We know that the ratio of copper area ($A_c$) in window to the total window area ($A_w$) is window space factor, which is given by

$$K_w = \frac{\text{copper area in window}}{\text{total window area}} = \frac{A_c}{A_w} \tag{2.10}$$

$$A_c = K_w A_w \tag{2.11}$$

Equating, Eqs. (2.9) and (2.11), we get

$$\frac{2AT}{\delta} = K_w A_w \tag{2.12}$$

From Eq. (2.2a),

$$Q = E_p I_p \times 10^{-3} = E_t \times T_p \times I_p \times 10^{-3} \tag{2.13}$$

[where volt/turn,

$$\left. E_t = \frac{E_p}{T_p} \Rightarrow E_p = E_t \times T_p \right] \tag{2.14}$$

Substituting $T_p I_p = \text{AT}$ from Eq. (2.8) in Eq. (2.13), we get

$$Q = E_t \times \text{AT} \times 10^{-3}$$

Substituting Eq. (2.12) in the above equation, we get

$$Q = E_t \times \frac{\delta K_w A_w}{2} \times 10^{-3}$$

Substituting $E_p/T_p$ from Eq. (2.3) in the above equation, we get

$$\begin{aligned} Q &= 4.44 \times f \times \phi_m \times \frac{\delta K_w A_w}{2} \times 10^{-3} \\ &= 2.22 f \varphi_m \delta K_w A_w \times 10^{-3} \end{aligned} \tag{2.15}$$

Substituting $\varphi_m = B_m$ (maximum flux density) $\times$ $A_i$ (area of core or area of iron) in the above equation, we get

$$Q = 2.22 f B_m \delta K_w A_w A_i \times 10^{-3} \tag{2.16}$$

Equation (2.16) is the output equation of a single-phase core-type transformer, and it relates the output kVA rating to its dimensions and parameters.

### 2.3.3 *Design of Single-Phase Shell-Type Transformer*

The cross-section of the core of the single-phase shell-type transformer is shown in Fig. 2.3. The arrangement of winding is shown in Fig. 2.4.

Here, both the windings are wound over the central limb and so each window has all the primary and secondary windings.

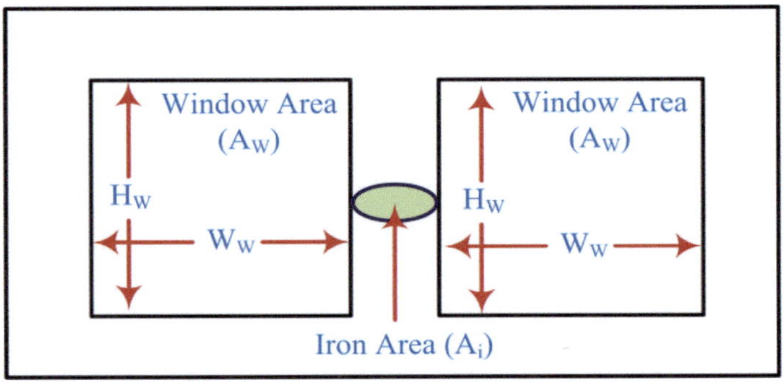


**Fig. 2.3** Cross-section of 1$\varphi$ shell-type transformer

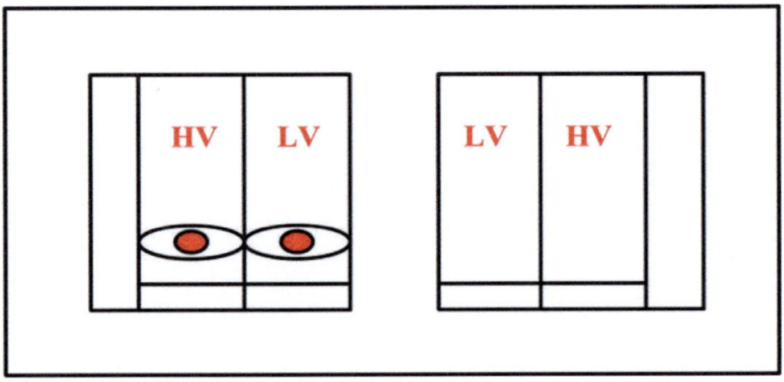


**Fig. 2.4** Winding arrangement of 1$\varphi$ shell-type transformer

Hence, the total copper area in the window is given by

$$A_c = T_pA_p + T_sA_s = \frac{2IT}{\delta} = \frac{2\text{AT}}{\delta} = K_wA_w$$

The aforementioned equation is the same as Eqs. (2.9) and (2.11), and so the output equation for a 1$\varphi$ shell-type transformer is the same as the output equation of a single-phase core-type transformer and is given by

$$Q = 2.22fB_m\delta K_wA_wA_i \times 10^{-3}.$$

The above equation is the output equation of a single-phase shell-type transformer.

## 2.3.4 *Design of Three-Phase Transformer*

The kVA rating of a three-phase transformer is given by

$$Q(\text{kVA}) = 3V_pI_p \times 10^{-3} \text{ (or ) } 3V_sI_s \times 10^{-3} \tag{2.17}$$

$$\simeq 3E_pI_p \times 10^{-3}[\because V_p \simeq E_p] \tag{2.18a}$$

$$\simeq 3E_sI_s \times 10^{-3}[\because V_s \simeq E_s] \tag{2.18b}$$

where p—primary, $V$, $E$—voltage, s—secondary, $I$—current.

### 2.3.5 Design of Three-Phase Core-Type Transformer

The cross-section of a three-phase core-type transformer is shown in Fig. 2.5. The arrangement of winding is shown in Fig. 2.6.

Each limb carries both the primary and secondary windings of one phase. So, in each window, there are two primary and two secondary windings. There are two windows, and in each window, the total copper area in the window is given by

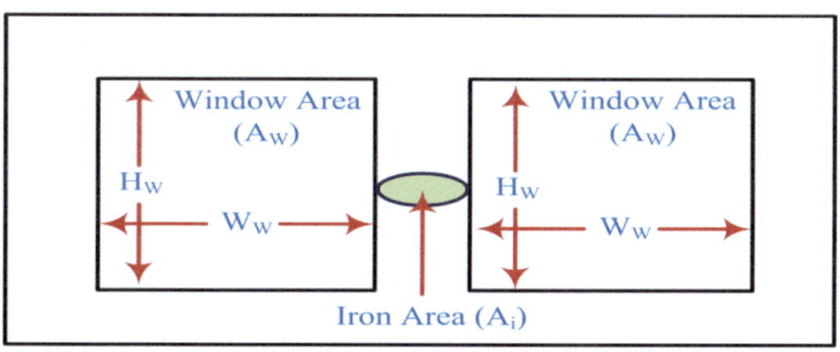


**Fig. 2.5** Cross-section of 3$\varphi$ core-type transformer

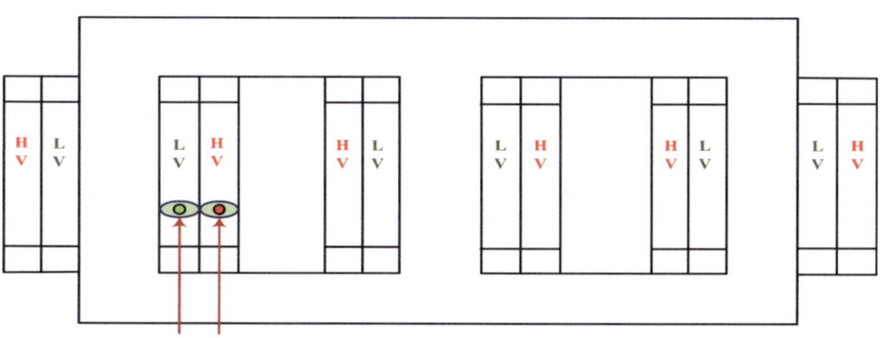


**Fig. 2.6** Winding arrangement of 3$\varphi$ core-type transformer

$$A_c = 2T_pA_p + 2T_sA_s \tag{2.19}$$

Following the same steps as used in the derivation of the output equation of a single-phase transformer, substituting Eq. (2.6) in Eq. (2.19), we get

$$A_c = 2T_p\frac{I_p}{\delta} + 2T_s\frac{I_s}{\delta} = \frac{2}{\delta}(T_PI_P + T_SI_S) \tag{2.20}$$

Substituting Eq. (2.8) in Eq. (2.20), we get

$$A_c = \frac{2}{\delta}(\mathrm{AT} + \mathrm{AT}) = \frac{4\mathrm{AT}}{\delta} \tag{2.21}$$

Similarly, equating Eqs. (2.21) and (2.11), we get

$$\frac{4\mathrm{AT}}{\delta} = K_wA_w$$

$$\mathrm{AT} = \frac{\delta K_wA_w}{4}. \tag{2.22}$$

From Eq. (2.18a),

$$Q = 3E_pI_p \times 10^{-3}$$

Substituting Eq. (2.14) in the above equation, we get

$$Q = 3E_t \times T_p \times I_p \times 10^{-3} \tag{2.23}$$

Substituting Eq. (2.8) in Eq. (2.23), we get

$$Q = 3E_t \times \mathrm{AT} \times 10^{-3}$$

Substituting Eq. (2.22) in the above equation, we get

$$Q = 3E_t \times \frac{\delta K_wA_w}{4} \times 10^{-3}$$

Substituting $E_p/T_p$ from Eq. (2.3) in the above equation, we get

$$Q = 3 \times 4.44 \times f \times \phi_m \times \frac{\delta K_wA_w}{4} \times 10^{-3}$$

$$Q = 3.33f\phi_m\delta K_wA_w \times 10^{-3} \tag{2.24}$$

Substituting $\varphi_m = B_m$ (maximum flux density) $\times\, A_i$ (area of core) in Eq. (2.24), we get

$$Q = 3.33 f B_m \delta K_w A_w A_i \times 10^{-3} \tag{2.25}$$

Equation (2.25) is the output equation of a three-phase core-type transformer.

### 2.3.6 Design of Three-Phase Shell-Type Transformer

The cross-section of a three-phase shell-type transformer is shown in Fig. 2.7. The arrangement of winding is shown in Fig. 2.8. This transformer has six windows and each window carries LV and HV windings of one phase only.

Copper area in each window is given by

$$A_c = T_p A_p + T_s A_s \tag{2.26}$$

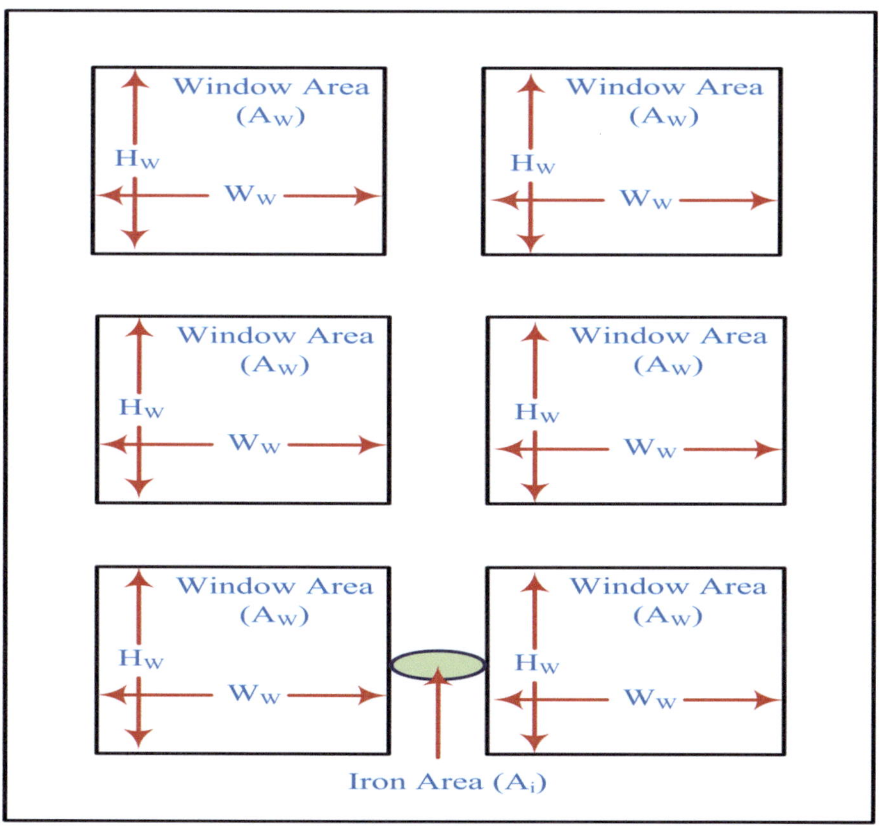


**Fig. 2.7** Cross-section of 3$\phi$ shell-type transformer

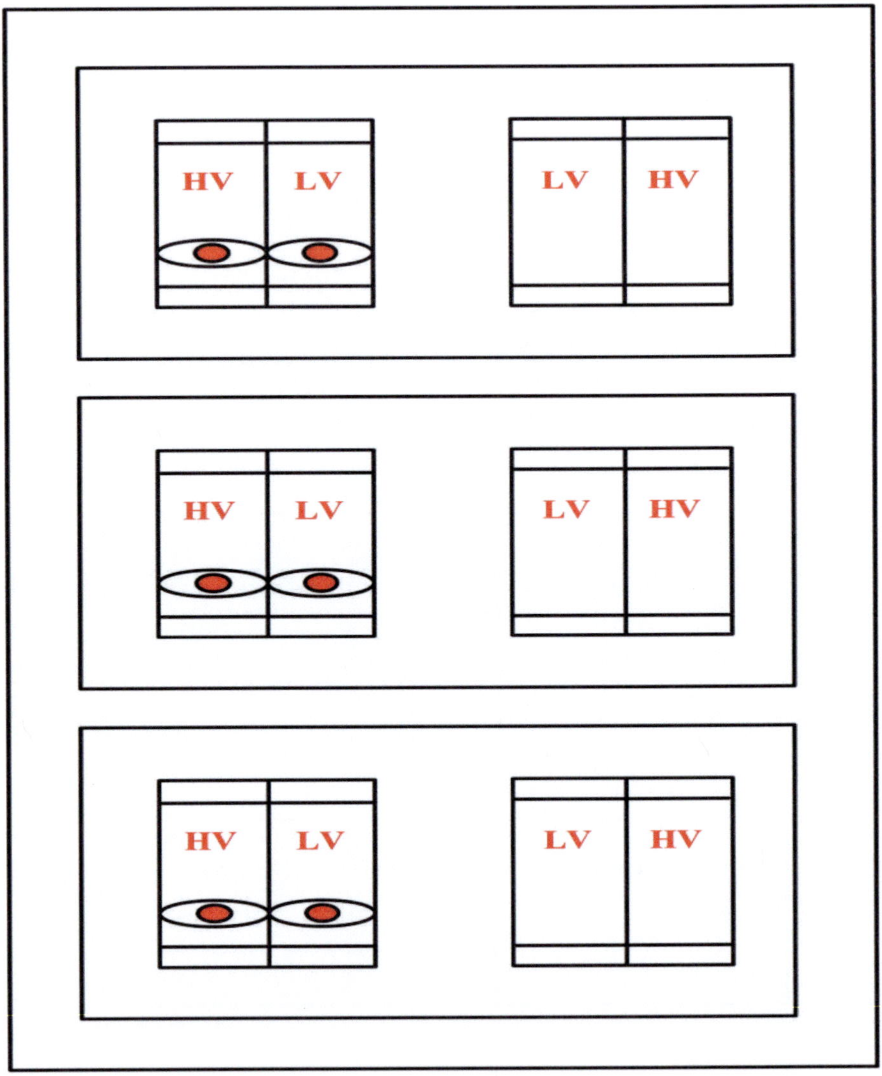


**Fig. 2.8** Winding arrangement of 3$\phi$ shell-type transformer

Following the same steps as used in the derivation of the output equation of a single-phase transformer, substituting Eq. (2.6) in Eq. (2.19), we get

$$A_c = \frac{T_p I_p}{\delta} + \frac{T_s I_s}{\delta}$$

$$A_c = \frac{T_p I_p + T_s I_s}{\delta} \tag{2.27}$$

Substituting Eq. (2.8) in Eq. (2.20), we get

$$A_c = \frac{\mathrm{AT} + \mathrm{AT}}{\delta}$$

$$A_c = \frac{2\mathrm{AT}}{\delta} \tag{2.28}$$

Similarly, equating Eqs. (2.21) and (2.11), we get

$$\frac{2\mathrm{AT}}{\delta} = K_w A_w$$

$$\mathrm{AT} = \frac{\delta K_w A_w}{2} \tag{2.29}$$

From Eq. (2.18a),

$$Q = 3E_p I_p \times 10^{-3}.$$

Substituting Eq. (2.14) in the above equation, we get

$$Q = 3E_t \times T_p \times I_p \times 10^{-3} \tag{2.30}$$

Substituting Eq. (2.8) in Eq. (2.30), we get

$$Q = 3E_t \times \mathrm{AT} \times 10^{-3}$$

Substituting Eq. (2.22) in the above equation, we get

$$Q = 3E_t \times \frac{\delta K_w A_w}{2} \times 10^{-3}$$

Substituting $E_p/T_p$ from Eq. (2.3) in the above equation, we get

$$Q = 3 \times 4.44 \times f \times \phi_m \times \frac{\delta K_w A_w}{2} \times 10^{-3}$$

$$Q = 6.66 f \phi_m \delta K_w A_w \times 10^{-3} \tag{2.31}$$

Substitution $\varphi_m = B_m$ (maximum flux density) $\times$ $A_i$ (area of core) in Eq. (2.31), we get

$$= 6.66 f B_m \delta K_w A_w A_i \times 10^{-3} \tag{2.32}$$

The above equation is the output equation of a three-phase shell-type transformer.

## 2.4 Volts per Turn of the Winding

The rating of a transformer per phase is given as

$$\begin{aligned} Q(\text{kVA}) &= VI \times 10^{-3} = 4.44f\phi_{\text{m}}TI \times 10^{-3} \\ &= 4.44f\phi_{\text{m}}\text{AT} \times 10^{-3} \qquad [\because TI = \text{AT}] \end{aligned} \tag{2.33}$$

The term, $\varphi_{\text{m}}$, used here, is called the magnetic loading, and the term AT is the electric loading.

This is a number, $\phi_{\text{m}} \equiv B_{\text{m}}A_{\text{i}}$, which is a measure of cross cross-section of the iron core due to magnetic loading. The efficiency of operation reduces with an increase in the magnetic loading and hence the flux density $B_{\text{m}}$, corresponding to which the core loss increases.

Similarly, the electric loading AT gives a measure of the cross-section of the winding. Increase in electric loading increases the number of turns and resistance and hence results in increased copper losses, reducing the efficiency.

So, higher values of $\varphi_{\text{m}}$ and AT will only reduce the efficiency of the transformer. For economical design of a transformer, the ratio of $\varphi_{\text{m}}$ and AT is maintained constant, which depends on the type of transformer.

The ratio,

$$x = \frac{\phi_{\text{m}}}{\text{AT}} \Rightarrow \text{AT} = \frac{\phi_{\text{m}}}{x} \tag{2.34}$$

Substituting Eq. (2.34) in Eq. (2.33), we get

$$\begin{aligned} Q &= 4.44f\frac{\phi_{\text{m}}^2}{x} \times 10^{-3} \\ \Rightarrow \phi_{\text{m}} &= \sqrt{\frac{x \times 10^3}{4.44f}} \times \sqrt{Q} \end{aligned}$$

Now, volt per turn of winding is given by

$$\begin{aligned} E_{\text{t}} &= 4.44f\phi_{\text{m}} \\ &= 4.44f\sqrt{\frac{x \times 10^3}{4.44f}} \times \sqrt{Q} \\ \Rightarrow E_{\text{t}} &= \sqrt{4.44fx \times 10^3} \times \sqrt{Q} \\ \Rightarrow E_{\text{t}} &= K\sqrt{Q} \\ \text{where } K &= \sqrt{4.44fx \times 10^3} \end{aligned}$$

**Table 2.1** Values of $K$ for various types of transformers

| Type of transformers | $K$ |
|---|---|
| Single-phase core type | 0.75–0.85 |
| Three-phase core type (power) | 0.6–0.7 |
| Three-phase core type (distribution) | 0.45 |
| Single-phase shell type | 1.0–1.2 |
| Three-phase shell type | 1.3 |

Here, $x$ and hence $K$ are constants. Their value depends on type, construction, and service conditions. Table 2.1 gives the various values of $K$ for different types of transformers.

## 2.5 Section of Flux Density

The EMF equation and the output equation indicate that the core area $A_i$ depends on flux density $B_m$. Lowering of the core area can be achieved by increasing the flux density. It will make the length of the mean turn of winding, which is only in proportion to the circumscribing circle diameter, possibly to take a small step back to shorten. Therefore, a massive amount of copper and iron will be saved. However, a savings of copper and iron with a higher flux density would also incur higher magnetizing current, leading to saturation as well as iron losses that would necessitate special cooler arrangements. The choice of $B_m$ also depends on the type and service conditions of the transformer. It has already been pointed out that a distribution transformer should be designed for lower iron losses, giving good all-day efficiency. Consequently, for distribution transformers, a lower flux density than in the cases considered in the above relation is assumed. Table 2.2 gives the permissible values of $B_m$ for different types of transformers. Flux density in a transformer also depends on the transformer's rating, with higher values of flux density corresponding to higher transformer ratings.

**Table 2.2** Values of flux density for various types of transformers

| Transformers' category | Flux density range |
|---|---|
| Hot-rolled silicon steel for distribution transformer | 1.1–1.4 $Wb/m^2$ |
| Hot-rolled silicon steel for power transformer | 1.2–1.5 $Wb/m^2$ |
| Cold-rolled grain-oriented steel for transformers up to 132 kV | 1.55 $Wb/m^2$ |
| For 275 kV | 1.6 $Wb/m^2$ |
| For 400 kV | 1.7 $Wb/m^2$ |

**Table 2.3** Values of current density for various types of transformers

| Type of transformers | Value of current density (a/mm$^2$) |
|---|---|
| Standard distribution transformer and small power transformer | 1.5–2.6 |
| Medium and large power transformer | 2.4–3.4 |
| Large power transformers | 5.4–6.3 |

## 2.6 Choice of Current Density

The choice of specific electric loading, i.e., the current density ($\delta$), will affect the transformer dimensions and efficiency. A higher value of current density will result in smaller dimensions of the transformer. But with increased electric loading, heating losses and hence efficiency decreases. Higher value of current density will result in excessive temperature rise, resulting in damage to the insulation due to local heating. Even with improved methods of cooling and with well-ventilated winding, the choice of current density is limited by the local heating. The type of transformer also affects the choice of current density, as the requirement of copper and iron losses is different for power and distribution transformers. Table 2.3 gives the range of permissible values of current density for different types of transformers.

## 2.7 Essentials of Designing the Core

The core forms the magnetic circuit of the transformer along with the yoke. The core is built up of silicon steel laminations of 0.3–0.5 mm thickness with varnish insulation in between.

The core section of a small-sized transformer is rectangular with circular or rectangular coils. For higher sizes, and adding advantage in terms of less space requirement with a reduction in copper consumption, it is preferred to select the rectangular core rather than the circular core and coils. It is also under consideration of the various force acts under the different fault situations. These mechanical forces can change the shape of the core section if the core is designed in a rectangular shape. Even, it is also able to make defects in the winding coil and its insulations. If we design the core in a circular shape, then such situations are avoided in the power system. Due to this reason, mostly all cores even laminations of the cores are designed in circular shape. Figure 2.9 shows the various shapes and laminations of cores for the core- and shell-shaped transformer.

As a laminated core needs to be used in a transformer, a larger number of laminations of different sizes are required to build a circular core. It also becomes difficult to arrange them together to get a circular section. So, it is a general practice to approximate a circular core as a stepped core with an infinite number of steps. However, the maximum number of steps can be restricted for practical reasons, with the minimum

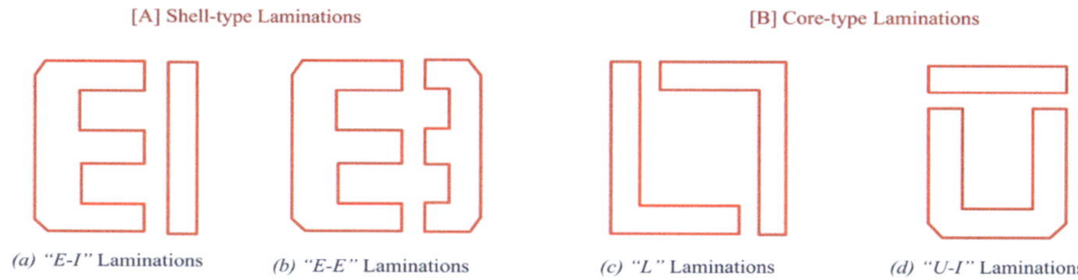


**Fig. 2.9** Shape of the various transformer cores and their laminations

**Fig. 2.10** Types of cores

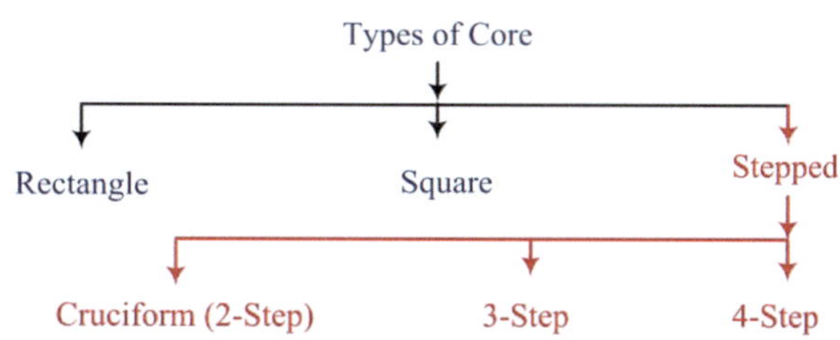


being one. A stepped core also enables the use of a circular coil. It is also called a cruciform type of core if the number of steps is two. Figure 2.10 gives the types of cores, and Fig. 2.11 shows the core sections of multi-stepped, square, and four-stepped cruciform cores. The shape of the core has to be selected based on minimum wastage of useful space and for a minimum length of copper. In power transformer design, the circumscribing circle takes the form of the smallest circle that will pass through an entire cross-section of the transformer core. The designers take advantage of the diameter of this circle to derive widths of the physical elements of stepped cores formed by assortments of rectangular laminations. Various dimensions are given in Fig. 2.11 to get the optimum cross-sectional area of the circumcircle with the number of steps.

Core is manufactured by combinations of stamping. Each stamping provides individual varnish called laminations. Varnish coating is very thin; however, the coating size is under consideration for manufacturing the core size. The complete core section is measured as the stack, which relates to the total cross-sectional area. It is defined

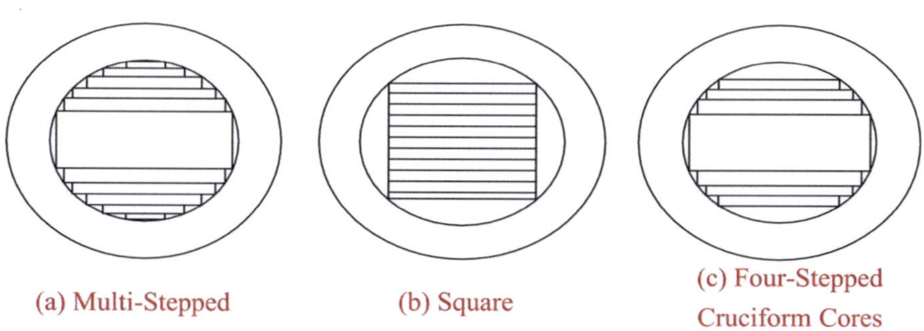


**Fig. 2.11** Core sections of **a** multi-stepped, **b** square, and **c** four-stepped cruciform cores

as,

$$\frac{\text{Net core section, } A_{\mathrm{i}}}{\text{Gross core section, } A_{\mathrm{gi}},} = Ks \text{ (Stacking factor).} \tag{2.35}$$

The stacking factor usually varies from 0.85 to 0.9.

In order to determine the utilization of the available space of the transformer core, the ratios, net core area to the circumscribing circle's area, and gross core area to the circumscribing circle's area are determined.

## 2.7.1 Square Core

The total area of the core under the square shape is elaborated in a cross-sectional view as per Fig. 2.12.

$$A_{\mathrm{gi}} = a \times a \tag{2.36a}$$

As per Fig. 2.12, $a$ is defined as a square side,

$$\text{Diagonal of Square} = \text{Diameter of Circumscribing} = d$$

Even in all squares,

$$d = \sqrt{2a}$$

Consequently,

$$a = \frac{d}{\sqrt{2}}d \tag{2.36b}$$

**Fig. 2.12** Square core

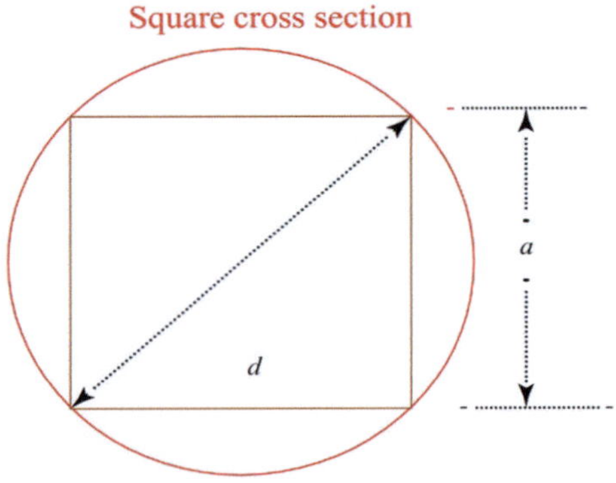

Replace with Eq. (2.36b) from Eq. (2.36a), we get

$$A_{\text{gi}} = \frac{d^2}{2} = 0.5d^2 \tag{2.36c}$$

To define the net core area,

$$A_i = \text{Stacking Factor} \times \text{Gross Core Area}$$

$$A_i = K_s \times A_{gi}$$

Suppose stacking factor = 0.9 and replace with Eq. (2.36c) in the aforementioned equation, we get

$$A_i = 0.9 \times 0.5d^2 = 0.45.$$

To find out the utilization space factor of the transformer from its accessible space, we have to obtain the circumscribing area of the circle, which is obtained as

$$A_{\text{cc}} = \frac{\pi d^2}{4}$$

Here, "$d$" is defined as the circle diameter.

Hence, the ratio of "$A_{\text{i}}$" (area of net core) to "$A_{\text{cc}}$" (area of circumscribing circles) is obtained as

$$\frac{\text{A}_{\text{i}}}{\text{A}_{\text{cc}}} = \frac{0.45d^2}{\frac{\pi d^2}{4}} = 0.58$$

And, the relation of "$A_{\text{gi}}$" gross core area to "$A_{\text{cc}}$" (area of circumscribing circles) is obtained as

$$\frac{A_{\text{gi}}}{A_{\text{cc}}} = \frac{0.5d^2}{\frac{\pi d^2}{4}} = 0.64$$

### 2.7.2 Stepped Core

The stepped core has two regions, region 1 and region 2, as shown in Fig. 2.13.

From Fig. 2.5, the gross core area is given by,

$$A_{\text{gi}} = \text{Area of Region 1} + \text{Area of Region 2}$$

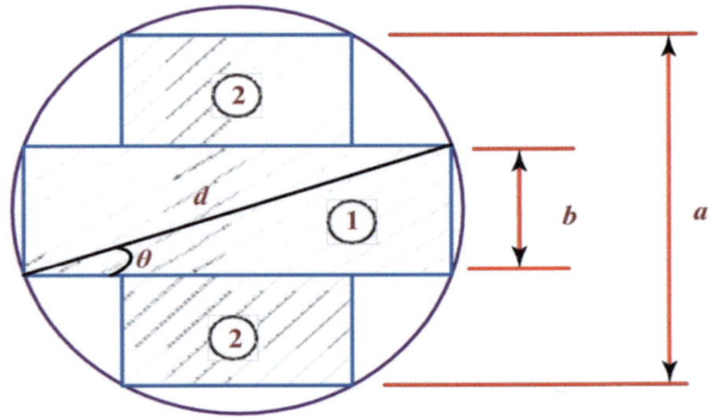


**Fig. 2.13** Stepped core

where

$$\text{Area of Region 1} = (a \times b)$$
$$\text{Area of Region 2} = (a - b) \times b$$

Hence,

$$A_{gi} = ab + ab - b^2$$

$$A_{gi} = 2ab - b^2 \tag{2.37}$$

It is required to express the terms "$a$" and "$b$" in terms of "$d$," which is the diameter of the circumscribing circle, to find the ratios, $\frac{A_i}{A_{cc}}$ and $\frac{A_{gi}}{A_{cc}}$.

From Fig. 2.13,

$$\cos\theta = \frac{a}{d} \quad \text{and} \quad \sin\theta = \frac{b}{d}$$
$$a = d\cos\theta \quad \text{and} \quad b = d\sin\theta \tag{2.38}$$

By adding Eq. (2.37) to Eq. (2.38),

$$A_{gi} = 2 \times d\sin\theta \times d\cos\theta - (d\sin\theta)^2$$
$$A_{gi} = 2d^2\sin\theta\cos\theta - d^2\sin^2\theta$$
$$A_{gi} = d^2\left|2\sin\theta\cos\theta - \sin^2\theta\right|, [\because 2\sin\theta\cos\theta = \sin 2\theta] \tag{2.39}$$

The maximum area for a given "$d$" is obtained when "$\theta$" is maximum. So, to get maximum "$\theta$," "$A_{gi}$" is differentiated with respect to "$\theta$" and equated to zero.

$\frac{dA_{gi}}{d\theta} = 0.$

Hence, on differentiating Eq. (2.6) concerning $\theta$ and connecting it to zero, we obtained

$$\frac{dA_{gi}}{d\theta} = d^2[2\cos 2\theta - 2\sin\theta\cos\theta] = 0$$
$$2\cos 2\theta = 2\sin\theta\cos\theta$$
$$2\cos 2\theta = \sin 2\theta$$
$$\frac{\sin 2\theta}{\cos 2\theta} = 2$$
$$\tan 2\theta = 2$$

$$2\theta = \tan^{-1}(2) \quad \text{and} \quad \theta = 31.72^{\circ} \tag{2.40}$$

Hence, with $\theta = 31.72°$, we get the maximum area of the core for a given $d$. Hence,

$$a = d\cos\theta = 0.85d, \quad \text{and} \quad b = d\sin\theta = 0.53d$$

Using Eq. (2.40) in Eq. (2.39), we get

$$A_{gi} = 0.618\,d^2$$

Net core area,

$$A_i = 0.9 \times A_{gi} = 0.56d^2$$

∵ stacking factor is assumed to be 0.9

$$\frac{A_i}{A_{cc}} = \frac{0.56d^2}{\frac{\pi d^2}{4}} = 0.71$$
$$\frac{A_{gi}}{A_{cc}} = \frac{0.618d^2}{\frac{\pi d^2}{4}} = 0.79$$

Hence, it is proved that the area of the circumscribing circle can be more effectively used by increasing the number of steps of the core. Table 2.4 shows the ratio of ($A_{gi}$/$A_{cc}$) and ($A_i$/$A_{cc}$) for various cores represented in Fig. 2.12 (Fig. 2.14).

**Table 2.4** ($A_{gi}/A_{cc}$) and ($A_i/A_{cc}$) for various cores

| Ratio | Square core | Two-stepped | Three-stepped | Four-stepped |
|---|---|---|---|---|
| $A_{gi}$/area of the circumscribing circle, $A_{cc}$ | 0.64 | 0.79 | 0.84 | 0.87 |
| $A_i$/area of the circumscribing circle, $A_{cc}$ | 0.58 | 0.71 | 0.75 | 0.78 |

**Fig. 2.14** Types of stepped cores

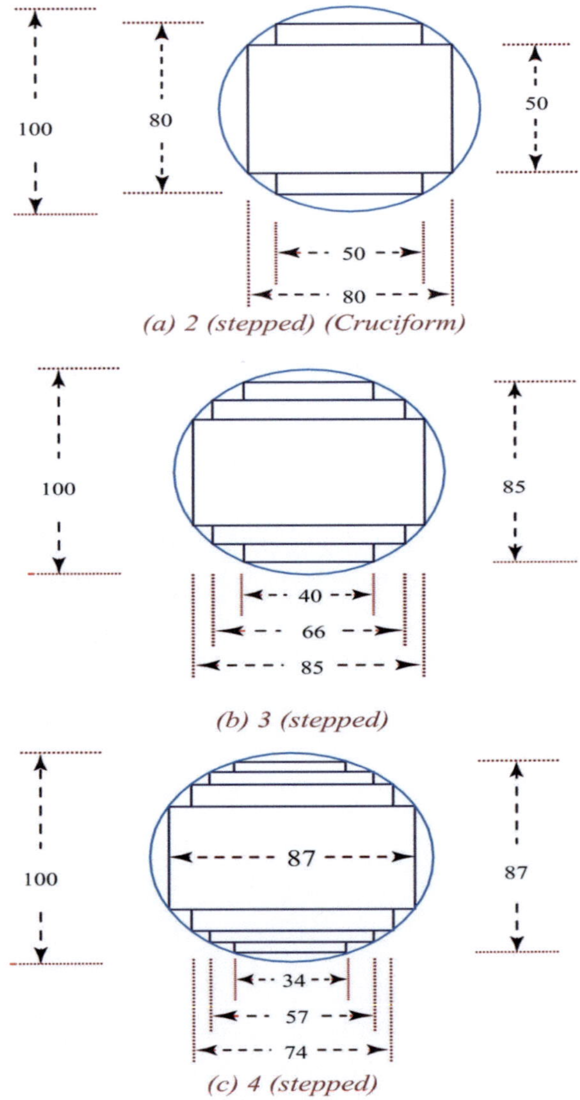


Note that the values in the second row of Table 2.1 can be obtained by multiplying the respective values in the first row. This is because $A_{\text{i}} = A_{\text{gi}} \times$ stacking factor. The stacking factor is assumed as 0.9.

## 2.8 Window and Core Proportions

The area of the window can be determined from,

$$\text{Area of window, } A_w = \frac{Q \text{ in kVA}}{2.22 \text{ or } 3.33 \text{ or } 6.66 f \delta B_m A_i K_w \times 10^3}. \tag{2.41}$$

If $W_W$ is the width of the window and $H_W$ is the window height, so, now, $A_W = W_W \times H_W$ to limit the transformer leakage reactance $H_W > W_W$. Normally ratio of $H_W/W_W$ track along the range of 2.5–3.5.

The core is the main part of the transformer, which completes the magnetic path with minimum leakage flux. Normally, the design of the core is based on its material and characteristics, which determine the magnetizing current and inrush current. If the core is not properly designed, then core losses increase and due to that, efficiency decreases in the transformer. Different ferromagnetic materials are used as a core material for stamping with proper lamination to reduce the core losses. However, the transformer losses are also affected by the design of the core. Optimizations of the core parameters have a major impact on the operating characteristics and reduction in losses in a transformer.

## 2.9 Design of Yoke

The yoke (horizontal section) of a transformer's core connects the legs (vertical section) and provides the least reluctance path. The windings are placed around the legs. The yoke portion does not have windings, and so the iron loss in the yoke can be minimized by increasing the yoke area, thereby reducing the operating flux density. Generally, the yoke area is made 20% more than the leg area. The yoke sections can be rectangular or cruciform. Figure 2.15 shows the yoke for rectangular and cruciform types of leg. A cruciform leg can also have a rectangular yoke as shown in Fig. 2.15.

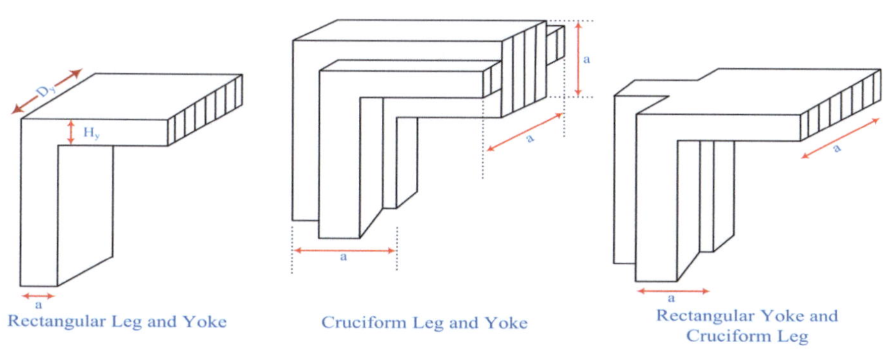


**Fig. 2.15** Shape of leg and yoke in the core assembly of a transformer

For a rectangular yoke, the depth $D_y$ of the yoke will be the same as the depth of the core. This will be equal to the width of the largest core stamping, "*a*." The height of the yoke $H_y$ can in general be taken as 1–1.5 times the width of the leg.

Area of yoke, $A_y = D_y \times H_y = 1.15\text{–}1.20$ times $A_{gi}$ (for hot-rolled silicon steel).

$A_y = A_{gi}$ (for cold-rolled silicon steel).

## 2.10 Conclusion

The design of a three-phase transformer is an intricate affair, as details of various electrical, magnetic, and mechanical parameters must be carefully considered for optimum efficiency, minimum losses, and reliable operation. The most prevalent structural design, which can be used in a transformer, is the core and shell type. Shell-type transformers are more effectively magnetic shielded and can withstand short-circuit current better and have a winding in the core of the transformer; core-type transformers with windings over separate limbs are easy the plaster and also good in cooling. Leakage reactance, total efficiency, and the connection of the flux in the windings depend on the design of both the core and shell types. Cylindrical windings are used in core-type transformers, while sandwich windings are used in shell-type transformers for improving magnetic coupling and lowering the reactance. As shown above, the volt per turn ratio is a principal design parameter for a transformer. For silicon steel cores, the flux density is usually selected between 1.2 and 1.7 Tesla and directly affects core losses and efficiency. The iron losses are increased with a higher flux density of the transformer, and a lower flux density leads to a bulkier transformer as a larger core is required. Therefore, the size of the winding conductors is similarly influenced by current density, specified in A/mm$^2$. This forces larger conductors to be used in order to reduce copper losses and may decrease efficiency because it results in a lower current density, whereas current density also becomes higher so that the heat generation becomes larger, causing thermal performance and the transformer life span to be degraded.

Transformer performance optimization is based on the core design. In order for the magnetic flux to be led to the core efficiently with a minimum amount of loss, laminated silicon steel is used to reduce eddy current losses in the core. Material utilization and efficiency are highly contingent on the selection of the core shape. A square core design is simple and easy to manufacture; however, this type of design results in greater flux leakage, while the use of a stepped core can improve the magnetic flux distribution, decrease loss, and optimize the utilization of material. Other requirements are that the proportions of the window and core have to be carefully balanced so that there is enough space for windings while taking up a compact size and good cooling. A good proportioned core will retain mechanical stability and improve efficiency in the transformer. Completing the magnetic circuit and structural support is important because of the yoke. Measures must be taken to reduce flux leakage and reluctance in the design, especially while maintaining adequate mechanical strength. To enable higher efficiency, usually laminated yokes are used to reduce eddy current loss. The flux distribution is desired to be uniform, and it is frequently desired to keep the core limb area of greater magnitude than the yoke's cross-sectional area to prevent saturation.

Therefore, the design of a three-phase transformer is a complicated procedure that involves carefully deciding the core structure, winding arrangement, electrical parameters, and mechanical considerations. The selection of configuration type, such as core type or shell type, is a function of application requirements, while the proper voltage per turn, flux density, and current density combination will maintain efficiency and reliability. Finally, core shape, window proportions, and yoke design are further optimized to maximize transformer performance.

## References


1. Saroha S, Shekher V, Kumar P, Kumar S (2021) Review of transformer core and winding design with material used. In: Pandey C, Goyat V, Goel S (eds) Advances in materials and mechanical engineering. Springer, Singapore, pp 39–49
2. Rahman Kiran M, Rabiul Islam M, Muttaqi KM, Sutanto D, Raad R (2024) A comprehensive review of advanced core materials-based high-frequency magnetic links used in emerging power converter applications. IEEE Access 12:107769–107799. https://doi.org/10.1109/ACCESS.2024.3436562
3. Xiong H et al (2024) Identification winding material of distribution transformer based on multi-information fusion. IEEE Trans Appl Supercond 34(8):1–5. https://doi.org/10.1109/TASC.2024.3456559
4. Valkovic Z, Rezic A (1992) Improvement of transformer core magnetic properties using the step-lap design. J Magn Magn Mater 112(1–3):413–415. https://doi.org/10.1016/0304-8853(92)91216-G
5. Hernández I, De León F, Cañedo JM, Olivares-Galván JC (2010) Modelling transformer core joints using Gaussian models for the magnetic flux density and permeability. IET Electr Power Appl 4(9):761–771. https://doi.org/10.1049/iet-epa.2010.0070
6. Rymar SV (2005) Optimization of a transformer with developed yoke magnetic stray fluxes. Avtom Svarka 7:32–35
7. More D, Jadhav D, Gaikwad S (2018) Parametric design to optimize coil clamping ring in power transformers using FEA. In: 2018 international conference on electrical, electronics, communication, computer, and optimization techniques (ICEECCOT), pp 834–839. https://doi.org/10.1109/ICEECCOT43722.2018.9001346
8. Hernandez C, Lara J, Arjona MA, Melgoza-Vazquez E (2023) Multi-objective electromagnetic design optimization of a power transformer using 3D finite element analysis, response surface methodology, and the third generation non-sorting genetic algorithm. Energies 16(5). https://doi.org/10.3390/en16052248
9. Transformer design optimization. In: Spotlight on modern transformer design. Springer, London, pp 331–376. https://doi.org/10.1007/978-1-84882-667-0_7
10. Adly A, Abd-El-Hafiz S (2025) Basics of power transformers design: rating-dimensions correlations. In: Unconventional performance oriented power transformers design methodologies. Springer, Cham, pp 29–49. https://doi.org/10.1007/978-3-031-85221-3_4
11. Zhao D, Wang Y, Jia Y, Du Y, Zhu Q, Zhang B (2024) Finite element analysis of physical properties of new transformer core materials. In: Dong X, Cai L (eds) The proceedings of 2023 4th international symposium on insulation and discharge computation for power equipment (IDCOMPU2023). Springer, Singapore, pp 407–416

12. Pramono WB, Wijaya FD, Hadi SP, Wahyudi MS, Indarto A (2023) Designing power transformer using particle swarm optimization with respect to transformer noise, weight, and losses. Designs 7(1). https://doi.org/10.3390/designs7010031
13. Phaengkieo D, Ruangsinchaiwanich S (2015) Design optimization of electrical transformer using artificial intelligence techniques. In: 2015 18th international conference on electrical machines and systems (ICEMS), pp 1381–1385. https://doi.org/10.1109/ICEMS.2015.7385254
14. Pamuk N (2022) Improving core quality in power distribution transformers using machine learning methods. Eurasia Proc Sci Technol Eng Math 21:46–54. https://doi.org/10.55549/epstem.1224559
15. Yi Z, Sun K, Liu H, Zhang Q (2023) Modeling and evaluation of ferrite eddy losses in high-frequency transformers based on loss-effective conductivity extraction. IEEE J Emerg Sel Top Power Electron 11(6):5990–6004. https://doi.org/10.1109/JESTPE.2023.3314062
16. Li FC et al (2019) Amorphous–nanocrystalline alloys: fabrication, properties, and applications. Mater Today Adv 4:100027. https://doi.org/10.1016/J.MTADV.2019.100027
17. Chiriac H, Lupu N (2005) Design and preparation of new soft magnetic amorphous ferromagnets. In: Idzikowski B, Švec P, Miglierini M (eds) Properties and applications of nanocrystalline alloys from amorphous precursors. Springer, Dordrecht, pp 165–176
18. Chothani N, Patel DD, Desai A (2023) Hysteresis current control-based boost rectifier analysis for single phase solid state transformer. In: 2023 international conference on recent advances in electrical, electronics & digital healthcare technologies (REEDCON), pp 385–390. https://doi.org/10.1109/REEDCON57544.2023.10151262
19. Patel D, Desai A, Chothani N (2023) Fixed frequency current control-based boost rectifier analysis for single phase solid state transformer. In: 2023 international conference for advancement in technology (ICONAT), pp 1–6. https://doi.org/10.1109/ICONAT57137.2023.10080301
20. Patel DD, Chothani N, Parmar BD (2023) Design and development of solid state transformer with different current control techniques. In: 2023 IEEE 11th region 10 humanitarian technology conference (R10-HTC), pp 414–419. https://doi.org/10.1109/R10-HTC57504.2023.10461788
21. Chothani N, Raichura M, Patel D (2023) Introduction to magnetic inrush of power transformer. In: Advancement in power transformer infrastructure and digital protection. Springer, Singapore, pp 71–99. https://doi.org/10.1007/978-981-99-3870-4_3
22. Raichura M, Chothani N, Patel D (2021) Review of methodologies used for detection of magnetising inrush and fault conditions in power transformer. IET Energy Syst Integr 3:109–129. https://doi.org/10.1049/esi2.12012
23. Chothani N, Raichura M, Patel D (2023) Adaptive biased differential protection considering over-fluxing and CT saturation conditions. In: Advancement in power transformer infrastructure and digital protection. Springer, Singapore, pp 187–229. https://doi.org/10.1007/978-981-99-3870-4_7
24. Chothani N, Patel D (2024) Modern method to discriminate high resistance internal winding faults and inrush condition/external faults in power transformer. In: 2024 3rd Odisha international conference on electrical power engineering, communication and computing technology (ODICON), 2024, pp 1–6. https://doi.org/10.1109/ODICON62106.2024.10797598
25. Raichura M, Chothani N, Patel D (2020) Development of an adaptive differential protection scheme for transformer during current transformer saturation and over-fluxing condition. Int Trans Electr Energy Syst 31:1–19. https://doi.org/10.1002/2050-7038.12751
26. Patel D, Chothani N (2020) Digital protective schemes for power transformer, 1st edn. In: Power systems. Springer, Singapore [Online]. Available at: https://books.google.co.in/books?id=tjaSzQEACAAJ
27. Chothani N, Patel D (2024) Phasor comparison and differential current-based earth fault detection for low-impedance grounded transformer in microgrid. In: Panda G, Basu M, Siano P, Affijulla S (eds) Proceedings of third international symposium on sustainable energy and technological advancements. Springer, Singapore, pp 275–285

# Chapter 3
# Determination of Main Dimensions of the Transformer with Numerical Solutions

In this chapter, the main dimensions of transformer windings are determined by applying the essential principles and methodologies. Finally, it provides a complete design approach taking into consideration both core type and shell type of single-phase and three-phase transformers. Geometries of core, including total height ($H$), total width ($W$), and the distance between adjacent limbs ($D$), are discussed in the context of what influence it has on the winding configuration. The chapter discusses critical dimensions like the diameter circumscribing circle ($d$), window width ($W_w$), window height ($H_w$), yoke height ($H_y$), and yoke depth ($D_y$) as these are important for the performance, cooling and mechanical strength of the winding assembly. The review includes various winding types (cylindrical, helical, continuous disk, and layer windings) and emphasizes construction, applications, and suitability for voltage level and transformer rating. Electrical, thermal as well as mechanical stresses are analyzed based on the choice of winding. The chapter brings out the fact that proper dimensioning is important for proper insulation coordination and loss minimization. Detailed solutions for numerically computed cases are given, illustrating practical design methods for determining most of the main parameters. Finally, this chapter offers a foundation for the design of safe and economically advantageous winding that is a prerequisite for transformer reliability and cost-effectiveness as well as operational efficiency.

## 3.1 Introduction

After involving magnetization in the field of electrical engineering, transformer is the key portion of a power transfer device [1–3]. The said device is an electrically isolated but magnetically coupled element.

Harmonics are the main cause of the losses, even though they also affect the protective solutions. A novel strategy is identified for the improvement of the transformer

 
N. Chothani et al., *Advancement in Infrastructure and Design Technology of Power Transformer*, Studies in Infrastructure and Control,
https://doi.org/10.1007/978-981-96-9420-4_3

efficiency with consideration of the harmonics. This concept totally involves an industrial transformer-based theme based on impedance matching strategy [4]. Different functionalities and different operation requirements are basic needs for the industry. Pulse transformers having high-power ratings with rapid discharge characteristic is one of the industry needed transformer application. The design of such type of a transformer was validated on the FEM-based software and also tested its key parameters on a prototype with satisfaction for > 500 kV, 100 kW, and 25 Hz frequency [5]. Current transformers (CT) differ from conventional transformers having drastic changes in the primary and secondary ratings in terms of the current and voltage. As far as concerns with sensing of the current signals having a measurement of the partial discharge (PD), high-frequency measurement is the preliminary requirement. After getting these issues, high-frequency CT is involved with cost-effective functionality [6, 7].

During a short circuit, heavy currents flow through the transformer winding. If the fault appears at the transformer's terminal, short-circuit value of the fault current drastically increases. An increase in the fault current causes different forces on the winding. Such kind of electromagnetic forces causes mechanical stresses on the winding. Based on the simulations, such a kind of analysis was carried out to analyze the transformer [8]. Even the position of the winding tape also affects the electromagnetic force [9]. Said forces are accounted for on the 1500 kVA transformer while considering the different forces, like radial and axial forces on different tap positions. Having several experimental tests, it is defined as the side-tapping and upper-lower tapping faces minimum force under different short-circuit conditions.

Transformer is designed with different vector groups as per the requirement of the power system and its applications. As far as the distribution transformer, one study was carried out for the analysis of the economic aspect of up to 100 kVA transformer in Bulgaria [10]. Ventilation also affects the working of a transformer, so environmental optimization is also an important factor for the indoor substation. The said study belongs to theoretical computation and different simulation testing with some safety feature limitations for 220 kV substations placed in indoor substations [11]. Under the high-frequency operation of a transformer, total losses increased per unit volume. So, for the reduction of losses, high-frequency transformers are involved in the field reduction of the size and losses with thermal designing consisting of forced air cooling [12–15]. The Tesla transformer is a novel invention of electrical engineering. The primary winding of the said transformer is proposed with the composite material of the combination of copper and titanium [16].

Winding capacitance and per-turn voltage distribution measurement is one of the important tasks. Impedance based on driving point per turn voltage distribution and capacitance was measured successfully with the help of the FEM technique [17]. Transformer dynamic psychoanalysis is highly significant in analysis, such as transient conditions or steady-state mode of operation in a power system [18]. During parallel operation of the transformers AC resistance and leakage inductance play major roles in operations [19] for high-frequency transformers. The dielectric design has a major impact on the operational parameters of the transformer. In a toroidal transformer dielectric design is adopted with optimum parameters with considerations of overvoltage due to transients and different dielectric stresses [20].

Winding design is the main factor for transformer optimization. Solid State Transformer (SST) is designed for different frequency analysis. The design of active parts like winding, insulation, and cooling systems is described in depth. All tests are carried out on AC and DC sources [21]. Even for medium frequency transformer (MFT), copper pipes are used as a winding with a liquid cooling [22] which gives better results than the normal transformer for power transfer and cooling purposes both. To create a parallel integrated resonant transformer that would be utilized with a high-current the emphasis should be on incorporating series resonant inductor (Lr) and parallel inductor (Lm) into the magnetic structure of the transformer [23]. Even with agricultural installations, transformer design varies with the features required of single phasing and unbalanced loading conditions. Monitoring and protection needs to be provided against such single-phasing and unbalancing conditions of the distribution transformer [24]. Even CNN [25, 26] and RVM [27, 28]-based regression techniques are also involved for the transformer protection. Superconducting material is also preferred for designing a transformer as a special device to reduce power loss and provide higher efficiency [29]. Transformer leakage impedance values are defined by the manufacturer, as per the consumer's demand. It is also preferable to optimize the gap position in transformer winding to improve leakage impedance as per specific requirements [30]. Multi-objective optimization technique is employed for the design of specific transformers with considerations of various parameters. A nanocrystalline material is preferred for this type of transformer [31]. Especially for hermetically sealed distribution transformers, multi-objective design optimization techniques are employed by considering the impact of current harmonics [32]. Said algorithm claims to be adaptively for all types of transformers to design and optimize the winding parameters, standards, and specifications. Thermal analysis provides a vital space for transformer design. Numerical and practical analysis of thermal analysis is required to design the transformer [33]. Thermal effect indicates the transformer via winding temperature indicator and oil temperature indicator.

The said chapter is divided into four sections. The first section elaborates on the importance and introduction to the transformer winding. The second portion of the chapter shows the detailed design of the winding parameters in a transformer. The third and fourth sections explain the different windings with their corresponding voltage rating. The fifth section involves different numerals that are involved with its solution and defines the main dimensions of the transformer, having a single or three-phase rating. Finally, the conclusion of the entire chapter is included, followed by the relevant references.

## 3.2 Computation of the Overall Dimensions

After designing the core, yoke sections, and the window sections, the overall dimensions could be computed as follows.

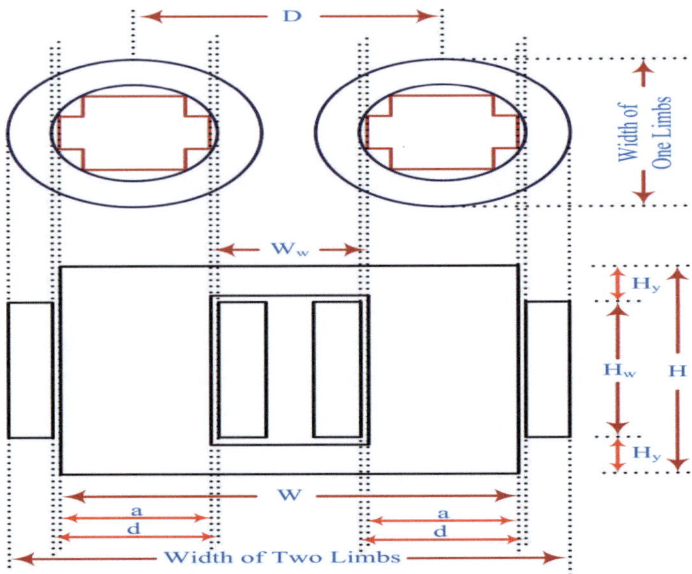


**Fig. 3.1** Representations of overall dimensions in 1-∅ core-type transformer

### 3.2.1 Single-Phase Core-Type Transformer

Figure 3.1 shows the overall dimensions of a single-phase core-type transformer, where $H$—overall height, $W$—overall width, $D$—distance between the adjacent limbs, $d$—diameter of circumscribing circle, $W_{\mathrm{w}}$—width of the window, $H_{\mathrm{w}}$—height of the window, $H_{\mathrm{y}}$—height of yoke, $D_{\mathrm{y}}$—depth of yoke.

$$H = H_{\mathrm{w}} + 2H_{\mathrm{y}}, \qquad D = W_{\mathrm{w}} + a, \qquad W = D + a, \qquad D_{\mathrm{y}} = a.$$

### 3.2.2 Three-Phase Core-Type Transformer

Figure 3.2 shows the overall dimensions of a three-phase core-type transformer, where

$$D = W_{\mathrm{w}} + d, \qquad H = H_{\mathrm{w}} + H_{\mathrm{y}}, \qquad W = 2D + a.$$

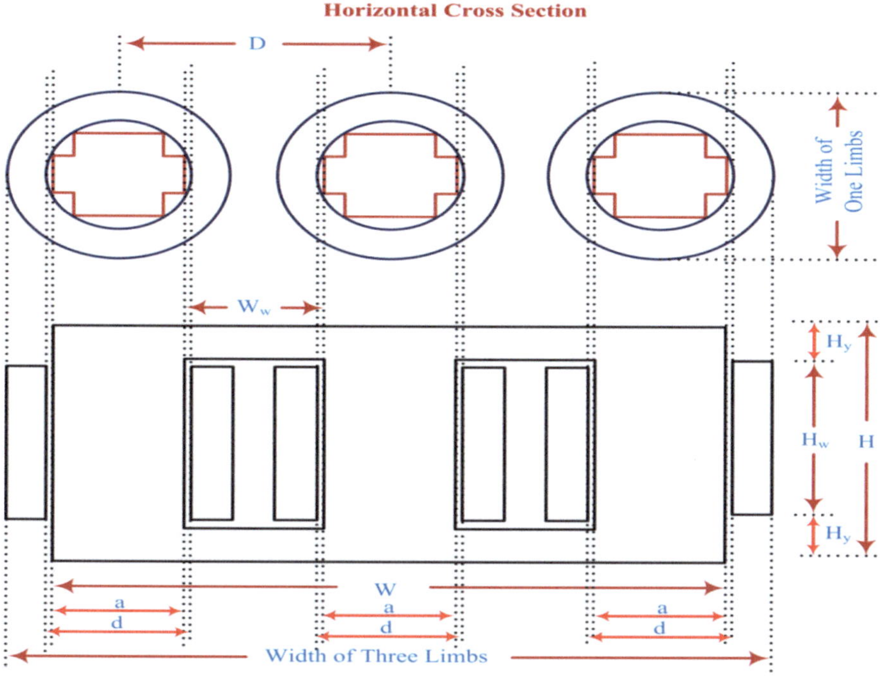


**Fig. 3.2** Representations of overall dimensions in 3-∅ core-type transformer

### *3.2.3 Single-Phase Shell-Type Transformer*

In the shell-type transformer shown in Fig. 3.3, the core section will be rectangular in shape. Only the central limb carries windings, and so the flux in the central limb is twice that of the outer limbs. Hence, the area of cross-section of the central limb is twice that of the other limbs.

The dimensions are given by

Gross core section of the central limb $= 2a \times b$.

| | |
|---|---|
| Over all width, | $W = 2W_w + 4a.$ |
| Overall height, | $H = H_w + 2H_y = H_w + 2a.$ |

*Note* The ratio $b/2a = 2.5$

### *3.2.4 Three-Phase Shell-Type Transformer*

Figure 3.4 shows the overall dimensions of a three-phase core-type transformer. The dimensions are given by

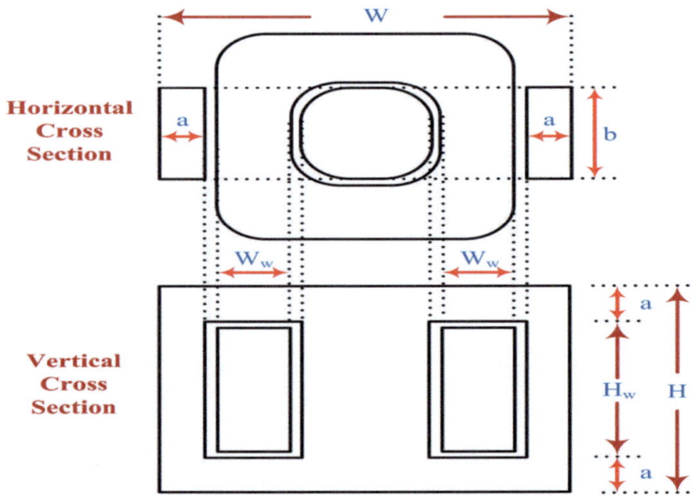


**Fig. 3.3** Representations of overall dimensions in 1-∅ shell-type transformer

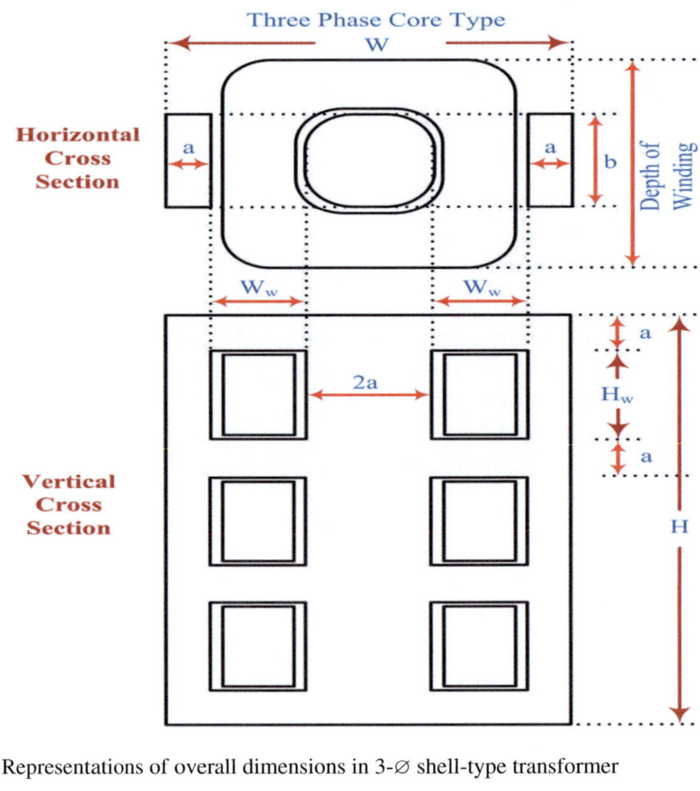


**Fig. 3.4** Representations of overall dimensions in 3-∅ shell-type transformer

Gross core section of the central limb = $2a \times b$.

| | |
|---|---|
| Over all width, | $W = 2W_w + 4a$. |
| Overall height, | $H = H_w + 2H_y = H_w + 2a$. |

## 3.3 Design of Windings

Normally, it is considered as whatever primary voltage $V_1$ is given on the transformer equal to the induced voltage $E_1$, let's consider as,

$$V_1 = E_1 = 4.44\varphi_m f T_1 = E_t T_1. \tag{3.1}$$

The ratio of turns and phase in the primary side or the number of turns in the primary is defined as

$$T_1 = \frac{V_1}{E_t}(\text{if the transformer is considered as a single - phase}), \tag{3.2}$$

$$= \frac{V_{1ph}}{E_t}(\text{for } 3 - \text{Ø transformers}). \tag{3.3}$$

The ratio of turns and phase on the secondary side, or the number of turns on the secondary, is defined as

$$T_2 = \frac{V_2}{E_t}(\text{for } 1 - \text{Ø transformers}), \tag{3.4}$$

$$= \frac{V_{2ph}}{E_t}(\text{for } 3 - \text{Ø transformers}). \tag{3.5}$$

To define current in the primary side or per phase current calculation, use the following terms for single-phase and three-phase transformers:

$$I_1 = \frac{\text{kVA} \times 10^3}{V_1}(\text{for } 1 - \text{Ø transformers}), \tag{3.6}$$

$$I_1 = \frac{\text{kVA} \times 10^3}{3 \times V_1}(\text{for } 3 - \text{Ø transformers}). \tag{3.7}$$

On the primary side, conductor's cross-sectional area is defined as

$$a_1 = \frac{I_1}{\delta}\text{mm}^2. \tag{3.8}$$

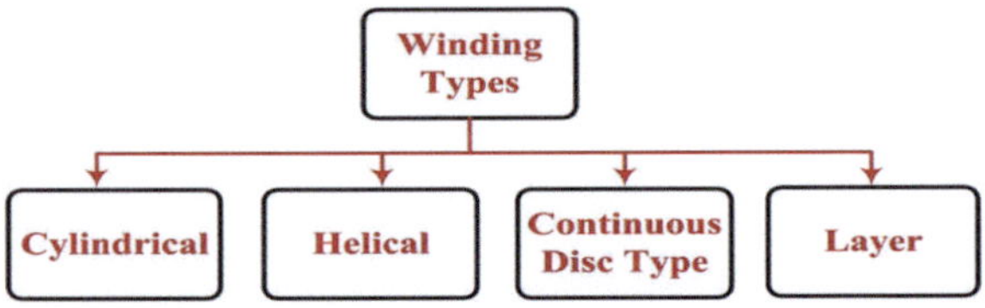


**Fig. 3.5** Different types of windings used in transformers

To define current in the secondary side or per phase current calculation, use the following terms for single-phase and three-phase transformers:

$$I_2 = \frac{\text{kVA} \times 10^3}{V_2} (\text{for } 1 - \text{Ø transformers}), \tag{3.9}$$

$$I_2 = \frac{\text{kVA} \times 10^3}{3 \times V_2} (\text{for } 3 - \text{Ø transformers}). \tag{3.10}$$

On the secondary side, conductor's cross-sectional area is defined as

$$a_2 = \frac{I_2}{\delta} \text{mm}^2. \tag{3.11}$$

Once the parameters of primary and secondary winding are calculated, such as currents, turns, and the cross–sectional area, it is easy to describe $H_\text{w}$ (window height), for those numbers of conductors with their window width ($W_\text{w}$).

The various types of windings used in transformers are shown in Fig. 3.5.

## 3.4 Types and Choice of Winding

Different types of windings, namely, cylindrical, helical, layered, and continuous disk windings, are generally used depending upon the voltage and current ratings. The various types of windings frequently used in transformer construction are explained as follows.

### *3.4.1 Cylindrical Winding*

Normally, cylindrical winding is placed with many layers; even there various types of conductors are used, like circular, strip, or rectangular shape. The layered cylindrical winding with sheet conductors is shown in Fig. 3.6a.

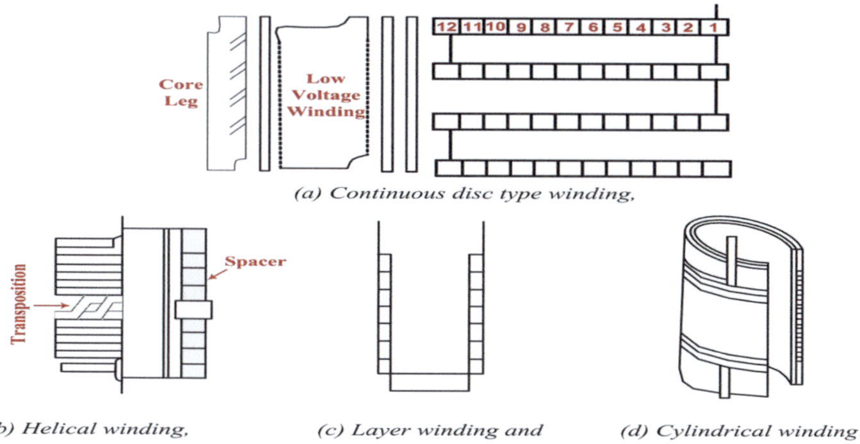


**Fig. 3.6** Different types of windings used in transformers: **a** Continuous disk-type winding, **b** helical winding, **c** layer winding, and **d** cylindrical winding

### 3.4.2 Helical Winding

Usually, these types of windings are used in low-voltage and high-current applications. Mostly, the number of turns is lesser, due to these high-current applications are fulfilled by these types of winding. Helical winding is mostly in single or double type which is wound on the cylinder of Bakelite. The axial direction is preferred for the turn wounding in single-layer helical winding. Different strands combined and connected in parallel called double-layer helical winding. Even, the number of turns in the helical windings is always lesser. Helical windings are primarily used in transformers for high-current, low-voltage applications. To increase the rating of current in helical winding, numbers of conductor strips are connected in parallel and also arranged besides each other with radial direction. These parallel-connected conductors can be transposed at regular intervals to equalize the resistance and leakage reactance of each conductor. A single helical winding is shown in Fig. 3.6b.

### 3.4.3 Continuous Disk Winding

For the high rating of the transformer, these types of windings are used. Series and parallel connections of disks or coils are connected as per the demand of the current and voltage ratings. Normally, disks are made by several parallel strips or individual strips wounded spirally on the core. Figure 3.6c shows an arrangement of a simple disk winding.

### 3.4.4 Layer Winding

Higher than the 100 kV range of the transformer, these types of windings are preferable. Numbers of layers are decided as per the voltage ratings of a transformer. To reduce the internal capacitance or scatter homogeneously of the winding capacitance, the outer layers are made shorter. Figure 3.6d shows an arrangement of layer-type winding. Normally, they are concentrically wound with series connections as shown in Fig. 3.6d.

After completing the selection of the appropriate winding as per the requirement of the transformer, it is suggested to find out the further dimensions of the window size, like ducts, size of proper insulations, and also clearance.

Table 3.1 gives a guideline for the choice of windings based on the voltage ratings.

## 3.5 Voltage Rating-Based Selection of Windings in Transformers

Transformer windings are chosen based on the voltage rating and power capacity of the windings, with the proper insulation, handling current, and mechanical strength to consider. Foil windings are also used at low voltages (3.3 kV) and small kVA ratings due to their small size and low cost. Helical windings are used in high-current, low-to-medium voltage applications because of superior cooling and strength. At medium voltages of around 11–33 kV, crossover types (or helical types) are used to aid harmonization of electrical stress. Beyond higher voltages, 33 kV and above, disc windings predominate since they are effective at controlling voltage distribution and may resist short-circuit forces. Interleaved or shielded disc windings provide impulse strength and insulation reliability at extra high voltages. The selection of a particular winding with respect to voltage rating and transformer power capacity is shown in Table 3.1.

## 3.6 Numerical with Solutions to Obtain the Main Parameters

***Example 3.1*** Core-type distribution transformer having a rating of 350 kVA, 3-Ø, 50 Hz, Y/Δ connected, 11,000/3300 voltage rating with an assumption that the distance between core centers is equal to double the core width. Identify the transformer's main dimensions.

**Solution**: Given,

Transformer rating, $Q = 350$ kVA, Type $= 3\emptyset$, $\mathrm{Y} - \Delta$ connection, Voltage ratio $= 11{,}000/3300$ V.

**Table 3.1** Voltage rating-based selection of windings in transformers

| Type of transformers | kVa | Voltage in kV | Winding |
|---|---|---|---|
| Transformers for rural supply | 1.6–100 | HV–3.3 to 11 | Foil, crossover |
| | | LV–0.44 | Helical |
| Distribution transformer, small power transformer | 100–1000 | HV–11 to 33 | Foil, crossover, disk, multi-layer helical |
| | | LV–0.44 | Helical, multi-layered helical |
| Transformers for secondary transmission | 10,000–20,000 | HV–33 to 66 | Disk |
| | | LV–11 | Disk or helix |
| Transformers for primary transmission | 45,000–15,00,00 | HV–132 to 440 | Disk or multi-layered helix |
| | | LV–11, 33, 66 | Disk or helix |
| Transformers at generating stations | 72,000–35,00,00 | HV–132 to 275 | Disk or multi-layered helix |
| | | LV–1122 | Disk or helix |

For volt per turn calculation, refer Table 2.1 in 3 − Ø distribution core-type transformer,

$$E_t = 0.45\sqrt{\text{kVA}} \tag{3.12}$$

$$E_t = 0.45\sqrt{350} = 8.4$$

$$\text{Ø}_m = \frac{E_t}{4.44\text{f}} = \frac{8.4}{4.44 \times 50} = 0.038\ \text{wb}$$

$$A_i = \frac{\text{Ø}_m}{\beta_m}(\text{Assuming 1.2T flux density})$$

$$A_i = \frac{0.038}{1.2} = 0.032\ \text{m}^2$$

For a Three-Stepped Core,

$$A_i = 0.6\text{d}^2 = 0.032\ \text{m}^2$$

$$d = \sqrt{\frac{0.032}{0.6}} = 0.23\ \text{m}$$

Largest Width of Stamping,

$$\text{a} = 0.9\text{d} = 0.9 \times 0.23 = 0.21\ \text{m}$$

Depth of yoke,

$$D_y = \text{a} = 0.21\ \text{m}$$

We know that,

$$\text{kVA} = 3.33 \times f \times \delta \times A_{\text{i}} \times B_{\text{m}} \times A_{\text{w}} \times k_{\text{w}} \times 10^{-3} \tag{3.13}$$

Let us assume $\delta$ as 2.5 A/mm$^2$
And,

$$k_{\text{w}} = \frac{10}{30 + \text{kV}_{\text{in}}} = \frac{10}{30 + 11} = 0.24$$

Substituting the determined values in Eq. (3.2), we get

$$350 = (3.33) \times (50) \times (0.032) \times (1.2) \times A_{\text{w}} \times (0.24) \times 10^{-3}$$
$$A_{\text{w}} = 0.09\ \text{m}^2$$

For calculation of the distance among the core center, $D = 2\text{a}$

$$W_{\text{w}} + d = 2a$$
$$W_{\text{w}} = 2a - d = 0.19\ \text{m}$$
$$H_{\text{w}} = \frac{A_{\text{w}}}{W_{\text{w}}} = \frac{0.09}{0.19} = 0.47\ \text{m}$$

To find out the overall width,

$$\begin{aligned} W &= 2D + a = 2(W_{\text{w}} + d) + a \\ &= 2(0.19 + 0.23) + 0.21 \\ &= 1.05\ \text{m} \end{aligned}$$

Overall height,

$$\text{H} = H_{\text{w}} + 2H_{\text{y}} = 0.89\ \text{m}$$
$$\text{Width} = 0.21\ \text{m}$$

Answer:

$$W = 1.05\ \text{m}, \quad H = 0.89\ \text{m}, \quad a = 0.21\ \text{m}$$

***Example 3.2*** Having transformer ratings for the $3 - \varnothing$, 50 Hz, $\Delta - Y$ connected transformer with 300 kVA and 6600/440 voltage ratios, find the main diminutions. The assumption for the core is three steps having 0.25 m diameter of a circumscribing circle with 0.4 m leg spacing, 8.5 V considered per turn having 2.5 A/ mm$^2$ current density with a window space factor and stacking factor are 0.28 and 0.9 concurrently. Also, determine the proportions of the conductor for the primary and secondary sides of the winding.

**Solution**: Given

Transformer rating = 300 kVA Volts/turn, $E_t = 8.5$ V
Voltage rating and type = 6600/440 V, $\Delta/Y$ Current density, $\delta = 2.5\ \text{A}/\text{mm}^2$

The diameter of the largest circle that can be drawn surrounding the core = 0.25 m.
Leg spacing = 0.4 m.
Window space factor, $k_w = 0.28$ Stacking factor, $k_i = 0.9$.

$$\text{Primary voltage} = 6000\ \text{V}(\Delta\ \text{connected}).$$

$$\text{Number of primary turns} = \frac{6600}{E_t} = \frac{6600}{8.5} = 776.$$

$$\text{Number of secondary turns} = \frac{\text{Secondary Phase Voltage}}{E_t} = \frac{\frac{440}{\sqrt{3}}}{8.5} = \frac{254}{8.3} = 30.$$

$$\text{Primary winding current} = I_{LP} = \frac{\text{kVA Rating of Transformer}}{\sqrt{3}V_{LP} \times 10^{-3}}$$
$$= \frac{300}{\sqrt{3} \times 6600 \times 10^{-3}} = 26.24\ \text{A}$$

Primary is $\Delta$ connected. So, the phase current of the primary $= \frac{26.24}{\sqrt{3}} =$ 15.15 A.For finding out the Area of primary winding conductors,

$$a_p = \frac{I_p}{\delta} = \frac{15.15}{2.5} = 6.06\ \text{mm}^2$$

Secondary Side Line Current $= \frac{Q}{\sqrt{3}V_{Ls} \times 10^{-3}}$
Since the secondary side is star-connected, phase current and line current are the same.

Cross-Section Area of the Secondary Conductor,

$$a_s = \frac{I_s}{\delta} = \frac{393.65}{2.5} = 157.5\ \text{mm}^2,$$

where $I_s = \frac{Q}{\sqrt{3}V_{Ls} \times 10^{-3}} = \frac{Q}{\sqrt{3} \times 400 \times 10^{-3}} = 393.65$ A
Copper area in the window,

$$A_c = 2(a_pT_p + a_sT_s) = 2[(6.06 \times 776) + (157.5 \times 30)] = 18855.12\ \text{mm}^2$$

Window area,

$$A_w = \frac{\text{Copper Area}}{\text{Window Space Factor}} = \frac{18855.12}{0.28} = 67339.71\ \text{mm}^2$$

Circumscribing circle area,

$$A_{cc} = \frac{\pi d^2}{4} = 0.049\ \text{m}^2$$

We know that for a three-stepped core,
Gross iron area,

$$\begin{aligned} A_{gi} &= 0.84 A_{cc} \\ &= (0.84)(0.049) \\ &= 0.041\ \text{m}^2. \end{aligned}$$

Net iron area,

$$\begin{aligned} A_i &= A_{gi} \times \text{stacking factor} \\ &= (0.041)(0.9) \\ &= 0.0369\ \text{m}^2. \end{aligned}$$

$$\text{Given leg spacing} = 0.45\ \text{m}(\because D = W_w + d = 0.45)$$

$$\begin{aligned} \text{Width of window } W_w &= D - d = 0.45 - 0.25 \\ &= 0.20\ \text{m} \end{aligned}$$

$$\text{Height of window} = H_w = \frac{A_w}{W_w} = \frac{0.067}{0.20} = 0.335\ \text{m}$$

Width of largest stamping, $a = 0.84\ d = 0.21\text{m}$
Total transformer's width ($W$),

$$\begin{aligned} W &= 2D + a \\ &= 2(0.45) + 0.21 \\ &= 1.11\ \text{m} \end{aligned}$$

Total transformer's height ($H$),

$$\begin{aligned} H &= 2(H_y) + H_w \\ &= 2(0.21) + 0.335 \\ &= 0.755\ \text{m} \end{aligned}$$

The main dimensions are

$$W = 1.11 \text{ m}, \qquad H = 0.755 \text{ m}, \qquad a = 0.21 \text{ m}$$

***Example 3.3*** A 3-Ø, 50 Hz, core-type transformer with oil cooled having dimensions as follows for two-stepped core:

Distance between the center of the two cores = 0.2 m.
Window height ($H_W$) = 0.24 m.
Circumscribing circle diameter = 0.14 m.
Core flux density =1.25 Wb/m$^2$.
Conductor current density =2.5 A/mm$^2$.
Consider a window space factor of 0.2 and a core area factor of 0.56.
Determine the kVA rating of a transformer.

**Solution**: Given

$$f = 50 \text{ Hz}, D = 0.2 \text{ m}, H_w = 0.24 \text{ m}, d = 0.14 \text{ m}$$

Core flux density = $B_m = 1.25 \text{ Wb/m}^2$
Conductor current density = $\delta = 2.5 \text{ A/mm}^2$.
Window space factor = $k_w = 0.2$.

$$\begin{aligned} Q &= \text{kVA rating of a transformer} \\ &= 3.33 \times \delta \times f \times A_i \times B_m \times k_w \times A_w \times 10^{-3} \end{aligned} \quad (3.14)$$

$$\begin{aligned} A_i &= 0.56 \text{ d}^2 = (0.56)(0.14)^2 \\ &= 0.0109 \text{ m}^2 \end{aligned}$$

For a 3 − Ø core-type transformer, $D = d + W_w$

$$\begin{aligned} W_w &= D - d = 0.2 - 0.14 \\ &= 0.06 \text{ m} \end{aligned}$$

$$\text{Area of window,} \quad \begin{aligned} A_w &= H_w \times W_w = 0.24 \times 0.06 \\ &= 0.0144 \text{ m}^2 \end{aligned}$$

$$Q = 3.33 \times 50 \times 1.25 \times 0.0109 \times 0.2 \times 0.0144 \times 2.5 \times 10^6 \times 10^{-3}$$

kVA rating of a transformer = $Q$ = 16.33 kVA.

***Example 3.4*** Calculate the core and window area of a 125 kVA, 2000/400 V, 50 Hz, 1Ø, shell-type transformer. Find also the amount of copper and iron required. Flux density is 1.1 T, current density is 2.2 A per mm$^2$, volts per turn given as 11.2 V, and window space factor is 0.33. Take the specific gravity of copper as 8.9 g/cm$^3$ and that of iron as 7.8 g/cm$^3$. Assume a rectangular core with a stamping of 7 cm width.

**Solution**: Given the kVA rating, $Q = 125$.
Volt and type of transformer = 2000/400 V 1Ø and
Shell flux density, $B_m = 1.1$ T.
Current density, $\delta = 2.2$ A per mm$^2$
Volts per turn = 11.2 V
Window space factor = 0.33 width of stamping, a = 7 cm.
Specific gravity of copper = 8.9 g/cm$^3$ Specific gravity of iron = 7.8 g/cm$^3$
Volts per turn, $E_t = 4.44 f Ø_m = 11.2\text{V}$

$$Ø_m = \frac{E_t}{4.44\text{f}} = \frac{11.2}{4.44 \times 50} = 0.05 \text{ wb}$$

$$A_i = \frac{Ø_m}{\beta_m} (\text{Shell Flux Density} = 1.1\text{T}) = \frac{0.05}{1.1} = 0.045 \text{ m}^2$$

We know that the kVA rating of a 1Ø shell-type transformer is given by,

$$Q = \text{kVA rating of a transformer} = 2.22 \times f \times \delta \times A_i \times B_m \times A_w \times k_w \times 10^{-3} \quad (3.15)$$

$$125 = (2.22) \times (50) \times (2.2 \times 10) \times (0.045) \times (1.1) \times A_w \times (0.33) \times 10^{-3}$$
$$A_w = 0.03 \text{ m}^2$$

The stampings are 7 cm in width. Hence,

$$a = 7 \text{ cm}$$
$$2a = 14 \text{ cm}$$

$$b = \frac{A_{gi}}{2a} = \frac{A_i}{k_s \times 2a}\left[\therefore A_{gi} = \frac{A_i}{k_s}\right]$$
$$= \frac{0.045}{0.9 \times 0.14} = 0.36 \text{ m}$$

Window area ($A_w$), $A_w = 0.03$ m$^2$.

Let the ratio of the height of the window to the width of the window be 3. The window is represented in Fig. 3.3 (Fig. 3.7).

$$\frac{H_w}{W_w} = 3$$

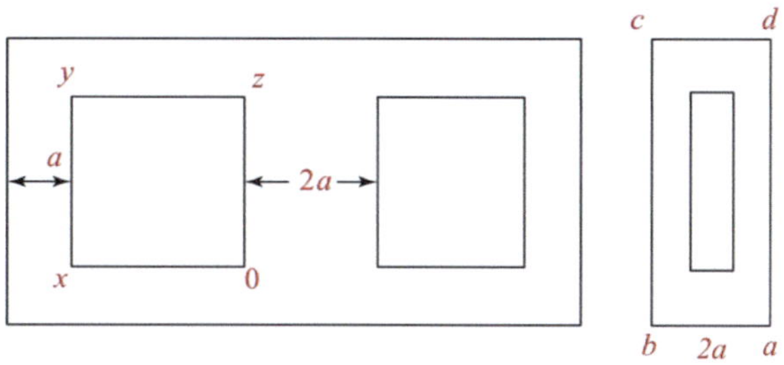


**Fig. 3.7** Window representation for Example 3.4

$$A_w = 0.03 = 3W_w^2$$

$$W_w = \sqrt{\frac{0.03}{3}} = 0.1\text{m}$$
$$H_w = 3 \times 0.1 = 0.3\text{m}$$

In this shell-type transformer, it is supposed that the whole window is occupied with LV as well as HV windings.

So, Winding height = Window height

And, LV and HV winding width = window width

$$\begin{aligned} \text{Copper's weight} &= \text{copper's volume} \times \text{density} \\ &= (\text{copper area} \times \text{copper length in bcdeb}) \times \text{density} \\ &= A_w k_w \times \text{length of mean turn} \times \text{density}, \end{aligned} \tag{3.16}$$

where,

$$\begin{aligned} \text{Length of mean turn of winding} &= 2(2a + W_w) + 2(b + W_w) \\ &= 2(14 + 10) + 2(32 + 10) \\ &= 132 \text{ cm}. \end{aligned}$$

From Eq. (3.2),

$$\begin{aligned} \text{Weight of copper} &= 0.03 \times 10^4 \times 0.33 \times 132 \times 8.9 \times 10^{-3} \\ &= 116.4 \text{ kg} \end{aligned}$$

$$\begin{aligned} \text{Weight of iron} &= 2 \times \text{volume of iron } \times \text{ density of iron} \\ &= 2 \times \frac{A_i}{2} \times \text{mean core length} \times \text{ density} \\ &= 2 \times \frac{0.045}{2} \times 10^4 \times [2(10 + 7) + 2(30 + 7)] \times 7.8 \times 10^{-3} \end{aligned}$$

Weight of iron = 379 kg.

***Example 3.5*** Discover the measurements of the core and window for a 125 kVA, 50 Hz, single-phase oil-immersed core-type transformer. Assume flux density as 1.23 T, current density as 2.5 A/mm$^2$, and window space factor as 0.32. Take EMF per turn as 17 V. Design a cruciform core with a window height of three times the window width.

**Solution**: Given

| | |
|---|---|
| Flux density, $B_m = 1.23$ T | Emf per turn = 17 V |
| Current density, $\delta = 2.5\ \text{A/mm}^2$ | Rating = 125 kVA |
| Window height, $H_w = 3W_w$ | Type = Single phase |

Window space factor, $k_w = 0.32$.
We know that
From Eq. (3.1),

$$\text{Volt per turn } E_t = 4.44 f\,\text{Ø}_m \tag{3.17}$$

Main flux,

$$\varphi_m = \frac{E_t}{4.44f} = \frac{17}{4.44 \times 50} = 0.0765\ \text{Wb}$$

Also, net iron area,

$$A_i = \frac{\text{Ø}_m}{\beta_m} = \frac{0.0765}{1.23} = 0.062\ \text{m}^2 \tag{3.18}$$

$$\text{The net iron area } (A_i)\text{for a cruciform core, } A_i = 0.56\,d^2 \tag{3.19}$$

From Eq. (3.3), the diameter of the circumscribing circle is given by

$$d = \frac{\sqrt{A_i}}{\sqrt{0.56}} = \frac{\sqrt{0.062}}{\sqrt{0.56}} = 0.332\ \text{m}$$

Largest stamping's width, $a = 0.85\text{d}$

$$a = (0.85) \times (0.332) = 0.282\ \text{m} \tag{3.20}$$

$\because$ Transformer's width = Largest stamping's width
Transformer's width = 0.282 m(app)

We know that

Smallest stamping's width, $b = 0.53d$

$$b = (0.53) \times (0.332) = 0.175\ \text{m} \tag{3.21}$$

Yoke height, $H_y = (1 \text{ to } 1.5)a$

$$H_y = (1)(0.282) = 0.282\ \text{m} \tag{3.22}$$

The kVA rating, $Q = 2.22f\ \delta A_i\ B_m A_w k_w \times 10^{-3}$

$$125 = (2.22)(50)(2.5 \times 10^6)(0.062)(1.23)A_w(0.32) \times 10^{-3} \tag{3.23}$$

$$A_w = \frac{125 \times 10^3}{6.77 \times 10^6} = 0.0184\ \text{m}^2$$

$$\because H_w = 3W_w \text{ and } A_w = H_w W_w \tag{3.24}$$

The main dimensions of the window and core are shown in Fig. 3.8.

$$A_w = 3W^2 = 0.0184$$

$$W_w = \sqrt{\frac{0.0184}{3}} = 0.078\ \text{m}$$

$$H_w = 3W_w = (3)(0.078) = 0.234\ \text{m}$$

Total transformer's height ($H$),

$$\begin{aligned} H &= H_w + 2H_y \\ &= 0.23 + 2(0.28) \\ &= 0.79\ \text{m} \end{aligned}$$

Overall transformer's width ($W$),

$$\begin{aligned} W &= W_w + d + a \\ &= 0.078 + 0.33 + 0.28 \\ &= 0.688\ \text{m} \end{aligned}$$

Total depth of transformer, $a = 0.282$ m
The main dimensions are

$$H = 0.79\ \text{m}, \quad W = 0.688\ \text{m}, \quad a = 0.282\ \text{m}.$$

***Example 3.6*** A 1-Ø core wounded transformer having 250 kVA, 50 Hz rating with 14Volts per turn through 1.2 Wb/m$^2$ maximum flux density. Determine the core and

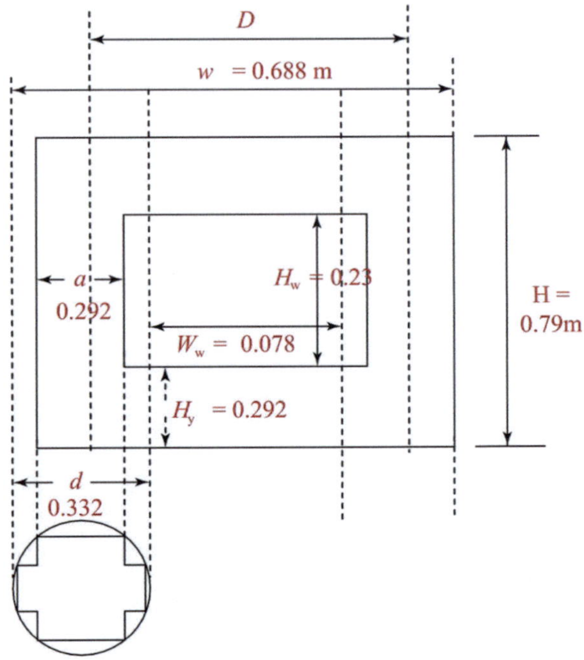


**Fig. 3.8** Main dimensions in window and core

yoke dimensions of the transformer. Consider 0.3 as a window space factor with 2.8 A/mm$^2$ current density. If the distance between the adjacent limbs is equal to 1.5 times the width of the transformer, design a square core. Assume a stacking factor of 0.8 and flux density in the yoke is 80% flux density in the core.

**Solution**:

Given,

kVA rating, $Q$ and type $= 250$ kVA and 1-Ø

Current density, $\delta = 2.8\ \text{A/mm}^2$.

$$\text{Frequency, } f = 50\ \text{Hz Distance between adjacent limbs, } D = 1.5a$$
$$\text{Volt/turn, } E_t = 14\ \text{V} \quad \text{Stacking factor} = 0.8$$

Maximum flux density, $B_m = 1.2\ \text{Wb/m}^2$ Flux density in yoke, $B_{my} = 80\%$ of $B_m$ in core

Window space factor, $k_w = 0.3$.

We know that

Volt/turn,

$$E_t = 4.44 \times f \times Ø_m$$
$$E_t = 4.44 \times f \times B_m A_i \tag{3.25}$$

From Eq. (3.1), area $A_i$ can be written as,

$$A_i = \frac{E_t}{4.44 \times f \times \beta_m} = \frac{14}{4.44 \times 50 \times 1.2} = 0.0525 \text{ m}^2$$

$$\text{For a square core,} \quad A_i = 0.45\, d^2$$

$$0.0525 = 0.45\, d^2 \tag{3.26}$$

$$d = \sqrt{\frac{0.0525}{0.45}} = 0.341 \text{ m}$$

Largest stamping's Width $(a)$, $a = 0.85\, d = 0.289$ m
Given the distance between adjacent limbs,

$$D = 1.5a = (1.5)(0.289)$$
$$= 0.4335 \text{ m}$$

$$\therefore \text{ Width of window, } W_w = D - d = 0.4335 - 0.341$$
$$= 0.0925 \text{ m}$$

For a one-phase transformer (core type),

$$Q = \text{Transformer's kVA rating} = 2.22 \times f \times \delta \times A_i \times B_m \times A_w \times k_w \times 10^{-3}$$
$$= 2.22 \times 50 \times 2.8 \times 10^6 \times 0.0525$$
$$\times 1.2 \times A_w \times 0.3 \times 10^{-3}$$
$$250 = 5874.12\, A_w \tag{3.27}$$

$$A_w = \frac{250}{5874.12} = 0.0425 \text{ m}^2$$

Height of window,

$$H_w = \frac{A_w}{W_w} = \frac{0.0425}{0.0925} = 0.4594$$

$$\text{Given that} \quad B_{myoke} = 80\% B_{mcore}$$
$$B_{my} = 0.8 B_{mc}$$

$$\frac{\varnothing_m}{A_y} = 0.8 \frac{\varnothing_m}{A_{gi}}$$
$$A_y = \frac{1}{0.8} A_{gi}$$
$$= 1.25 \times 0.0656 = 0.082 \text{ m}^2,$$

where

$$\frac{A_{\mathrm{i}}}{\text{Stacking Factor}} = A_{\mathrm{gi}} = \frac{A_{\mathrm{i}}}{0.8} = \frac{0.0525}{0.8} = 0.0656\ \mathrm{m}^2. \quad (3.28)$$

$$\begin{aligned} &\text{Depth of yoke, } D_{\mathrm{y}} = a = 0.289\ \mathrm{m} \\ &\text{Height of yoke } H_{\mathrm{y}} = \tfrac{0.082}{0.289} = 0.283\ \mathrm{m} \\ &\text{Overall width, } \begin{aligned} W &= D + a = 0.4335 + 0.289 \\ &= 0.7225\ \mathrm{m} \end{aligned} \\ &\text{Overall height, } \begin{aligned} H &= H_{\mathrm{w}} + 2H_{\mathrm{y}} \\ &= 0.4594 + (2)(0.283) \\ &= 1.0254\ \mathrm{m} \end{aligned} \end{aligned}$$

The main dimensions of the transformer are,

$$W = 0.72\ \mathrm{m}, \quad H = 1.0254\ \mathrm{m}, \quad a = 0.289\ \mathrm{m}.$$

***Example 3.7*** For a 200 kVA rating 3-Ø core-type transformer having a 6600/600 Voltage ratio, find the main dimensions. Assume a three-stepped, cruciform core with an area factor of 0.62 and window space factor of 0.25. The current density is 2.8 A/mm$^2$ and the emf per turn is 9 V. Assume the height of the window as three times the width of the window and the maximum flux density as 1.25 Wb/m$^2$.

**Solution**: Given

Rating, $Q$ and type = 200 kVA and 3Ø Maximum flux density = 1.25
Wb/m$^2$ Voltage = 6600/600 V Current density = 2.8 A/mm$^2$
Window space factor $k_{\mathrm{w}} = 0.25$ Volt/turn = 9V

$$\frac{H_w}{W_w} = 3$$

We know that,

$$\text{Volt per turn,} \quad E_{\mathrm{t}} = 4.44 f\ B_{\mathrm{m}} A_{\mathrm{i}} \quad (3.29)$$

$$\text{Net iron area,} \quad \begin{aligned} A_{\mathrm{i}} &= \frac{E_t}{4.44 \times f \times \beta_m} \\ &= 0.0324\ \mathrm{m}^2 \end{aligned}$$

$$\begin{aligned} \text{For a four - stepped core,} \quad A_{\mathrm{i}} &= 0.62\ \mathrm{d}^2 \\ 0.0324 &= 0.62\ \mathrm{d}^2 \end{aligned} \quad (3.30)$$

$$d = \sqrt{\frac{0.0324}{0.62}} = 0.2286 \text{ m}$$

Width of largest stamping for a four-stepped section,

$$a = 0.92 \text{ d}$$
$$\Rightarrow a = 0.210 \text{ m} \tag{3.31}$$

kVA of a 3Ø transformer is,

$$Q = \text{kVA rating of a transformer} = 3.33\delta f A_i B_m k_w A_w \times 10^{-3} \tag{3.32}$$

$$200 = (3.33)(50)(1.25)(0.0324)A_w(0.25)\left(2.8 \times 10^6\right) \times 10^{-3}$$

Window area,

$$A_w = \frac{200}{4.72 \times 10^3} = 0.0423 \text{ m}^2$$

Given that,

$$\frac{\text{Height of window}}{\text{Width of window}} = \frac{H_w}{W_w} = 3$$

Area,

$$A_w = (H_w)(W_w) = 0.0423 \text{ m}^2$$
$$A_w = 3W_w^2 = 0.0423$$
$$W_w = \sqrt{\frac{0.0423}{3}} = 0.11874$$
$$H_w = 0.1188 \times 3 = 0.3564 \text{ m}$$

We know that $H_y = a$ and $Dy = a$

$$\begin{aligned} H &= H_w + 2H_y \\ \text{Overall height,} \quad &= 0.3564 + 2(0.210) \\ &= 0.7764 \text{ m} \\ W &= 2D + a = 2W_w + 2d + a \\ \text{Overall width,} \quad &= (2)(0.1188) + 2(0.2286) + 0.210 \\ &= 0.9048 \text{ m} \end{aligned}$$

The main dimensions of the transformer are,

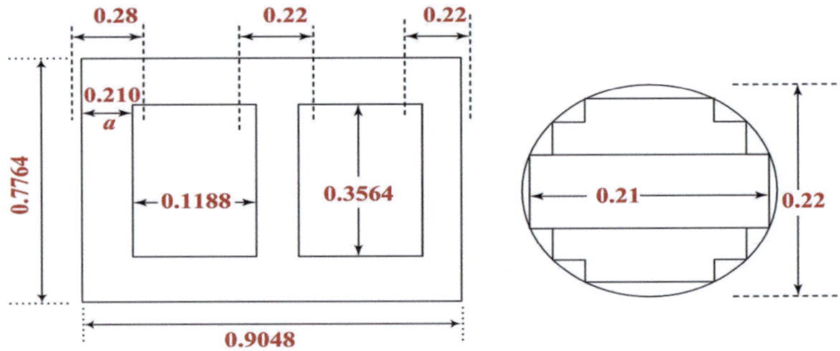


**Fig. 3.9** Main dimensions in window and core for Example 1.5.7

$$W = 0.9048 \text{ m}, \qquad H = 0.7764 \text{ m}, \qquad a = 0.210 \text{ m}.$$

The main dimensions of the window and core are shown in Fig. 3.9.

***Example 3.8*** Determine the main dimensions of a transformer with a rating of 100 kVA, 50 Hz, 1-Ø core type, and a voltage ratio of 2200/480. Compute the primary and secondary conductors' cross-sectional area and number of turns as well. Assume Volt/turn as 7.5 V, maximum flux density as 1.2 Wb/m$^2$. The window space factor is 0.28, the height-to-width ratio is 2, the current density is 2.5 A/mm$^2$, and the ratio of the effective cross-section area of the core to twice the diameter of the circumscribing circle is 0.6.

**Solution**:

Given Rating, $Q$ and type $=$ 100 kVA and 1Ø Current density, $\delta = 2.5\ \text{A/mm}^2$
Voltage $= 2200/480\text{V}$ Window space factor, $k_w = 0.28$
Volt/turn $= 7.5$ $A_i = 0.6d^2$

$$B_m = 1.2\ \text{Wb/m}^2$$

$$\frac{\text{Height of window}}{\text{Width of window}} = \frac{H_w}{W_w} = 2$$

Volt per turn, $E_t = 4.44 f B_m A_i = 7.5$
Net iron area,

$$A_i = \frac{E_t}{4.44 \times f \times B_m} = \frac{7.5}{4.44 \times 50 \times 1.2} = 0.02815\ \text{m}^2$$

$$\text{Given} \quad A_i = 0.6\ \text{d}^2$$

$$0.02815 = 0.6\ \text{d}^2$$

$$d = 0.2166 \text{ m} \tag{3.33}$$

For analysis purpose, let us assume a three-stepped core and the largest stamping's width,

$$a = 0.9d \tag{3.34}$$

$$\begin{aligned} a &= 0.1949 \\ Q &= 2.22f\, B_{\mathrm{m}} A_{\mathrm{i}} A_{\mathrm{w}} k_{\mathrm{w}} \delta \times 10^{-3} \\ 100 &= (2.22)(50)(1.2)(0.02815) A_{\mathrm{w}} (0.28)\left(2.5 \times 10^{6}\right) \times 10^{-3} \\ A_{\mathrm{w}} &= 0.03809 \text{ m}^2 \end{aligned}$$

Given

$$\frac{\text{Height of window}}{\text{Width of window}} = \frac{H_{\mathrm{w}}}{W_{\mathrm{w}}} = 2$$

and $A_{\mathrm{w}} = 0.03809 \text{ m}^2$

Hence,

$$\begin{aligned} (H_{\mathrm{w}})(W_{\mathrm{w}}) &= 0.03809 \\ 2W_{w}^{2} &= 0.03809 \\ W_{\mathrm{w}} &= 0.138 \text{ m} \\ H_{\mathrm{w}} &= 0.276 \text{ m} \\ \text{Depth of yoke} &= a = 0.1949 \text{ m} \end{aligned}$$

Let the area of the yoke is 20% greater than the area of the core section,

$$A_{\mathrm{y}} = 1.2\frac{A_{\mathrm{i}}}{0.9} = 1.2\frac{0.02815}{0.9} = 0.03783 \text{ m}^2$$

$$\because \quad D_{\mathrm{y}} = 0.1949 \text{ m}$$

and $A_{\mathrm{y}} = 0.03753 \text{ m}^2$

$$H_{\mathrm{y}} = \frac{A_{\mathrm{y}}}{D_{\mathrm{y}}} = \frac{0.03753}{0.1949} = 0.1925 \text{ m}$$

Overall width,
$$W = D + a = W_w + d + a$$
$$= 0.138 + 0.2166 + 0.1949$$
$$= 0.5495 \text{ m}$$

Overall height,
$$H = H_w + 2H_y = 0.276 + 2 \times 0.1925$$
$$= 0.661 \text{ m}$$

Number of secondary turns,

$$T_2 = \frac{\text{Secondary Voltage}}{\text{Volts Per Turn}} = \frac{480}{7.5} = 64$$

Number of primary turns,

$$T_1 = \frac{\text{Primary Voltage}}{\text{Volts Per Turn}} = \frac{2200}{7.5} = 293$$

Primary current,

$$I_1 = \frac{\text{kVA} \times 1000}{\text{Primary Voltage}} = \frac{100 \times 1000}{2200} = 45.45 \text{ A}$$

Area of a cross-section of the primary winding,

$$a_1 = \frac{I_1}{\delta} = \frac{45.45}{2.5} = 18.18 \text{ mm}^2$$

Secondary current,

$$I_2 = \frac{\text{kVA} \times 1000}{\text{Secondary Voltage}} = \frac{100 \times 1000}{480} = 208.34 \text{ A}$$

Area of a cross-section of the secondary winding,

$$a_2 = \frac{I_2}{\delta} = \frac{208.34}{2.5} = 83.34 \text{ mm}^2$$

## 3.7 Conclusion

Transformer winding dimensions' accurate determination is a key part of transformer design, as it determines the transformer's performance as well as its durability and cost. In this chapter, we presented a structured way to calculate such parameters as overall height, width, window's dimensions, and yoke dimensions. However, the accuracy of these elements is critical to the satisfactory accommodation of windings, for adequate insulation recesses, and for maintaining the structural integrity under operation. The single-phase and the three-phase configurations of both core-type and shell-type transformers are addressed. However, the structural layout of these types varies and so different decisions are required as to winding layout and size. For core-type transformers, windings are stacked concentrically around the limbs so that space utilization and cooling are of paramount importance. Since windings are enclosed in the core, it leads to more compact but robust structure in shell type. Further, the winding types covered were cylindrical, helical, continuous disk, and layer windings. For any given application, there are advantages of each type: for instance, helical windings with low voltage and high currents, layered or disk windings with high voltage and poor insulation management. It is shown that the selection of winding type is very sensitive to transformer voltage rating and application. Additionally, it includes posed, as well as solved, numerical problems, which give practical insight into real-world design scenarios. Some will therefore help reinforce theoretical concepts while showing how dimensional decisions affect the final transformer design.

## References


1. Chothani N, Raichura M, Patel D (2023) Transformer infrastructure for power grid. In: Advancement in power transformer infrastructure and digital protection. Springer Nature Singapore, Singapore, pp 1–26. https://doi.org/10.1007/978-981-99-3870-4_1
2. Patel D, Chothani N (2020) Digital protective schemes for power transformer, first. In: Power systems. Springer Singapore, Singapore. [Online]. Available: https://books.google.co.in/books?id=tjaSzQEACAAJ
3. Chothani N, Patel D, Raichura M (2019) Transformer protection with sequence components and digital filters, 1st edn. LAP LAMBERT Academic Publishing, Latvia, European Union
4. Liu Q, Liu F, Yang Y, Peng Y, Liu K-Z, Wang A (2024) A novel transformer winding design strategy for efficiency improvement and harmonic sup-pression in industrial power system. Prot Control Mod Power Syst 9(6):30–41. https://doi.org/10.23919/PCMP.2023.000302
5. Feng B, Fu Y, Dong L, Qin S, He M (2024) Design and characteristics of a high-power pulse transformer based on a closed cage core structure. IEEE Trans Plasma Sci 52(6):2241–2250. https://doi.org/10.1109/TPS.2024.3441320
6. Fritsch M, Wolter M (2022) High-frequency current transformer design and construction guide. IEEE Trans Instrum Meas 71:1–9. https://doi.org/10.1109/TIM.2022.3177189
7. Chothani N, Raichura M, Patel D (2023) Current transformer infrastructure and its application to power system protection. In: Advancement in power transformer infrastructure and digital protection. Springer Nature Singapore, Singapore, pp 101–133. https://doi.org/10.1007/978-981-99-3870-4_4
8. Gutten M, Janura R, Jurcik J (2015) Determination of the electromagnetic forces caused by short-circuit currents. In: 2015 56th international scientific conference on power and electrical engineering of Riga technical university (RTUCON), pp 1–4. https://doi.org/10.1109/RTUCON.2015.7343179
9. Dawood K, Bağrıyanık FG, Kömürgöz G (2024) Effect of tapping position on the electromagnetic forces of the tap-winding transformer. J Electr Eng Technol 19(8):4989–5000. https://doi.org/10.1007/s42835-024-01894-9
10. Atanasov V, Mechkov E (2024) Technical economic analysis concerning connection groups of distribution transformers in Bulgaria. In: 2024 23rd international symposium on electrical

apparatus and technologies (SIELA), pp 1–5. https://doi.org/10.1109/SIELA61056.2024.10637866

11. Zhu Y, Ma Y, Li Y, Yang H, He K, Zhang Q (2024) Design and optimization of ventilation environment in the main transformer room of 220 kV indoor substation. In: 2024 the 8th international conference on energy and environmental science (ICEES 2024), Y. Liu, Ed., Cham: Springer Nature Switzerland, pp 627–643
12. Ngo M, Cao Y, Dong D, Burgos R, Nguyen K, Ismail A (2023) Forced air-cooling thermal design methodology for high-density, high-frequency, and high-power planar transformers in 1U applications. IEEE J Emerg Sel Top Power Electron 11(2):2015–2028. https://doi.org/10.1109/JESTPE.2022.3223849
13. Chothani N, Patel DD, Desai A (2023) Hysteresis current control-based boost rectifier analysis for single phase solid state transformer. In: 2023 international conference on recent advances in electrical, electronics and digital healthcare technologies (REEDCON), pp 385–390. https://doi.org/10.1109/REEDCON57544.2023.10151262
14. Patel D, Desai A, Chothani N (2023) Fixed frequency current control-based boost rectifier analysis for single phase solid state transformer. In: 2023 international conference for advancement in technology (ICONAT), pp 1–6. https://doi.org/10.1109/ICONAT57137.2023.10080301
15. Patel DD, Chothani N, Parmar BD (2023) Design and development of solid state transformer with different current control techniques. In: 2023 IEEE 11th region 10 humanitarian technology conference (R10-HTC), pp 414–419. https://doi.org/10.1109/R10-HTC57504.2023.10461788
16. Zhao L, Li R, Zhang Y, Xu XD, Shang W, Qiu XD (2022) Design and test of a copper-titanium-composite primary winding for tesla transformer. IEEE Trans Plasma Sci 50(9):3155–3159. https://doi.org/10.1109/TPS.2022.3199215
17. Samal M, Lakshmi Prasanna K, Mishra P, Mondal M (2024) Accurate estimation of transformer winding capacitances and voltage distribution factor using driving point impedance measurements. IEEE Access 12:133670–133684. https://doi.org/10.1109/ACCESS.2024.3460968
18. Patel D, Shah S, Chothani N, Raichura M (2021) Dynamic psychoanalysis of DC separately excited machine on laboratory visualization. In: 2021 international conference on sustainable energy and future electric transportation (SEFET), pp 1–6. https://doi.org/10.1109/SeFet48154.2021.9375770
19. Pniak L, Quéval L, Revol B, Teu J-SN, Gautier C, Béthoux O (2023) AC resistance and leakage inductance estimation for planar transformers with parallel-connected windings. IEEE Trans Power Electron 38(1):728–738. https://doi.org/10.1109/TPEL.2022.3203515
20. Hussain MK, Gómez P (2023) Optimal dielectric design of medium voltage toroidal transformer with electrostatic shield under fast front excitation. IEEE Trans Power Deliv 38(2):1395–1405. https://doi.org/10.1109/TPWRD.2022.3214101
21. Cremasco A, Rothmund D, Milone M, Mokkapaty S, Lomonova EA (2024) 166 kW liquid-insulated medium-frequency transformer with hybrid foil-litz windings. IEEE Open J Power Electron 5:201–213. https://doi.org/10.1109/OJPEL.2024.3354807
22. Djekanovic N, Dujic D (2022) Copper pipes as medium frequency transformer windings. IEEE Access 10:109431–109445. https://doi.org/10.1109/ACCESS.2022.3214651
23. Yuan T, Jin F, Li Z, Zhao C, Li Q (2023) Design of an integrated transformer with parallel windings for a 30-kW LLC resonant converter. IEEE Trans Power Electron 38(11):14317–14333. https://doi.org/10.1109/TPEL.2023.3291954
24. Chothani N and Kumar S (2025) A New technique for monitoring and protection of distribution transformer during single-phasing situation. In: 2nd international conference on Sustainable Energy Technologies and Computational Intelligence (SETCOM). pp 1–6. https://doi.org/10.1109/SETCOM64758.2025.10932397
25. Raichura M, Chothani N, Patel D (2021) Efficient CNN-XGBoost technique for classification of power transformer internal faults against various abnormal conditions. IET Gener Transm Distrib 15(5):972–985. https://doi.org/10.1049/gtd2.12073

26. Chothani N, Raichura M, Patel D (2023) Convolution neural network and XGBoost-based fault identification in power transformer. In: Advancement in power transformer infrastructure and digital protection. Springer Nature Singapore, Singapore, pp 231–262. https://doi.org/10.1007/978-981-99-3870-4_8
27. Patel D, Chothani NG, Mistry KD, Raichura M (2018) Design and development of fault classification algorithm based on relevance vector machine for power transformer design and development of fault classification algorithm based on relevance vector machine for power transformer. IET Electr Power Appl 12(4):557–565. https://doi.org/10.1049/iet-epa.2017.0562
28. Dharmesh Patel NC. Relevance vector machine based transformer protection. In: Digital protective schemes for power transformer, 1st edn. Springer, Singapore, Springer, Singapore, pp 107–131
29. Shu A, Tian B, Chang S, Ma Y, Pi W (2024) Structure design of 500 kVA HTS transformer and Joule loss in superconducting windings. Phys C Supercond Appl 620:1354482. https://doi.org/10.1016/j.physc.2024.1354482
30. Dawood K, Tursun S (2024) Optimized gap positions for improved leakage impedance in dry-type transformer design. Results Eng 23:102632. https://doi.org/10.1016/j.rineng.2024.102632
31. Zhang P et al (2024) Research on multi-objective optimization method in the design of high-power nanocrystalline alloy high-frequency transformer. J Magn Magn Mater 606:172389. https://doi.org/10.1016/j.jmmm.2024.172389
32. Hashemi MH, Kilic U (2024) Multi-objective design optimization of hermetically sealed core-type distribution transformer considering current harmonics of power grid using NSGA III. Eng Sci Technol Int J 55:101745. https://doi.org/10.1016/j.jestch.2024.101745
33. Cetinceviz Y, Sehirli E (2024) Design and numerical analysis of thermal capability of power transformer using coupled electromagnetic field-thermal model. Electr Eng 106(4):4821–4834. https://doi.org/10.1007/s00202-024-02271-5

# Chapter 4
# Resistance and Leakage Reactance Estimation of HV/LV Winding

The resistance and leakage reactance of high-voltage (HV) and low-voltage (LV) windings are critical parameters that strongly impact the performance, efficiency, and stability, and therefore must be carefully monitored in power transformers. The real part of the transformer's impedance, winding resistance, is dependent on the factors of conductor material (usually copper or aluminum), cross-sectional geometry, operating temperature, and frequency. Other influences, such as skin and proximity effects at higher frequencies, increase effective resistance and must be corrected for using correction factors. The resistance is also affected by the temperature coefficient of the conductor; this process of temperature compensation is an important requirement for accurate resistance calculation. There is a component of impedance called leakage reactance, which is due to magnetic flux that does not link both windings and which passes through either one of sole windings. This property directly impacts the short-circuit behavior as well as voltage regulation. Winding geometry and spacing between HV and LV windings affect the level of leakage reactance, and this is influenced by the number of turns. Leakage flux and reactance are reduced when the windings are interleaved and increased by non-interleaved designs.

## 4.1 Introduction

Power transformers are very essential equipment in the electricity supply system for transforming higher voltage to lower voltage and vice versa [1, 2]. Essential design and calculation aspects that have the most impact on the performance of a transformer include the winding resistance and leakage reactance. All these parameters are essential for voltage control, power factor, short-circuit capacity, and performance stability of power systems and power system protection [3]. Even inrush generations also depend on these parameters [4].

 
N. Chothani et al., *Advancement in Infrastructure and Design Technology of Power Transformer*, Studies in Infrastructure and Control,
https://doi.org/10.1007/978-981-96-9420-4_4

Winding resistance and leakage reactance cause quite a stir in the transformer's behavior during various situations. Although resistance is associated with copper losses and heating, leakage reactance controls short-circuit impedance and dynamics. The winding resistance gives the actual, or resistive, part of the transformer's overall impedance. The resistance depends on the conductor material (for example copper or aluminum), the cross-sectional area and conductor geometry, and temperature variations with load and ambient conditions. It is also influenced by frequency-dependent phenomena such as skin and proximity effects. Leakage reactance, the second part of transformer impedance, is apparent when there is weak coupling between the two windings of the transformer. Its value depends on various factors such as the physical arrangement of windings, spacing, and insulation between HV and LV windings and the geometry and magnetic properties of the core and surrounding materials. From a design aspect, determining the amount of resistance and leakage reactance is always a compromise of cost, efficiency, and reliability [5]. Initial approximations of resistance and reactance are obtained from analytical models that employ Maxwell's equations and empirical formulas. Modern calculations, for example, finite element analysis, permit precise computations of electromagnetic fields for improving these estimates [6]. Theoretical calculations are confirmed through laboratory analysis of DC resistance measurement and short-circuit impedance tests. Various standards, including IEC 60,076 and IEEE C57, describe the ways to measure parameters with high accuracy [7, 8].

Recent research focuses on various tasks such as incorporating high-performance materials with superior electrical and thermal properties, utilizing machine learning algorithms to predict transformer parameters and developing new winding configurations to optimize reactance and resistance [9]. Resistance and reactance values are also affected by the generation of harmonics in the core during saturation. These harmonics also affect the protective schemes and give operation under abnormal conditions [10, 11]. Momentary overfluxing is also involved as a malfunctioning of the protective schemes [12, 13]. The power transformer is a vital equipment in the new network, which helps to change voltage levels in order to provide energy to various operational requirements. Proper estimation of resistance and leakage reactance in transformer windings is significant to impart improved performance, durability, and dependability. This research focuses on the overview and details of both resistance and reactance in and between HV and LV windings of power transformers with actual consideration from the facet of design [14]. Transformer windings are considered the generation of as main elements by which electrical energy is transmitted. Their reluctance and resonance are an essential part of specifying the electric characteristics of transformers and help to control the voltage, determine efficiency, and improve protection against faults. The copper losses are attributed to resistance, mainly of the winding material itself, whereas leakage reactance is due to imperfect magnetic coupling and is the basis of the short-circuit impendence of the transformer [15]. The determination of these parameters is a process that can be highly involved, and there is a need to understand material science, electromagnetic theory, and good practical modeling. Engineering concepts of resistance and reactance are applied to the transformer design and operation procedures through analysis, measurement, and computation. This introduction provides the concepts of resistance and leakage reactance in both HV and LV windings, including their importance, factors affecting them, and how to calculate them [16]. It also explores current literature and technology innovation in transformer designs and provides an understanding of new trends that improve performance, dependability, and the green future of power transformer systems [17]. Even now days there are trends for the high-frequency transformer to reduce the size of a transformer for the same rating power transformer [18–20]. Sometimes its functionality is also adopted via a solid-state transformer (SST).

## 4.2 Resistance Calculation

On calculating the resistances of the low-voltage and high-voltage windings, it is possible to estimate the copper loss in the transformer. The cooling system will be designed based on the losses that occurred and hence lead to a rise in the temperature [21]. The mean length of LV and HV windings is calculated from the cross-sectional area of coils, the number of turns, and the disposition of the coils. Then, the resistance of each winding can be calculated as follows:

Primary winding/phase resistance,

$$R_{\mathrm{p}} = \frac{\rho L_{\mathrm{mt}} T_{\mathrm{p}}}{a_{\mathrm{p}}}. \tag{4.1}$$

Secondary winding/phase resistance,

$$R_s = \frac{\rho L_{\mathrm{mt}} T_s}{a_s}. \tag{4.2}$$

Mean length of turn of the primary winding,

$$\mathrm{L_{mt}} = \pi \times \text{mean diameter of the winding}. \tag{4.3}$$

Mean length of turn of the secondary winding,

$$\mathrm{L_{mt}} = \pi \times \text{mean diameter of the winding}, \tag{4.4}$$

where resistivity of copper at 60 °C, $r = 2.1 \times 10 - 6$ ohm − cm or $2.1 \times 10^{-8}$ohm − meter or 0.021 ohm/m/mm$^2$.

$L_{\mathrm{mtp}}$ and $L_{\mathrm{mts}}$ are the lengths of mean turns of primary and secondary windings, respectively.

$T_{\mathrm{p}}$ and $T_{\mathrm{s}}$ represent the number of turns.

$a_{\mathrm{p}}$, $a_{\mathrm{s}}$ are the area of cross-sections of the conductors.

Total resistance per phase referred to primary side,

$$R_t = R_p + \left[\frac{T_p}{T_s}\right]^2 R_s. \tag{4.5}$$

## 4.3 Reactance Calculation

The reactance of the windings primarily involves the assessment of leakage flux. Reactance calculations are required to make an estimate of the regulation of the transformer to be designed.

(a) **Useful Flux**

The magnetic flux that links the primary and secondary windings is called the useful flux, by which the electromagnetic linkage transmits energy from the primary side to the secondary side. The path of the useful flux lies within the magnetic core.

(b) **Leakage Flux**

The leakage flux contributes to the inductance of windings and links with only the primary or secondary winding. The trajectory of the leakage flux is reliant upon the geometric arrangement of the coils and the adjacent iron masses.

(c) **Reactance**

1. Leakage reactance = $2\pi fl$ × inductance = $2\pi$f × (flux linkage per current).
2. Flux linkage = flux × number of turns.
3. Flux = (mmf or ampere turn)/reluctance = *Ampere Turn* × permeance ( ∧).
4. Permeance(∧) = 1/Reluctance = $\frac{a\mu_0\mu_r}{}$Permeance ( ∧ ) = $\frac{1}{\text{Reluctance}}$ = $\frac{a\mu_0\mu_r}{1}$, where

   *a* Area over which the flux is established.

   *l* Flux path length.

If the leakage reactance of the primary and secondary windings is denoted by $X_p$ and $X_s$, respectively, then the total leakage reactance on the primary winding will be given as

$$X_p = x_p + x_s\prime = x_p + x_s\left(\frac{T_p}{T_s}\right)^2. \tag{4.6}$$

Similarly, the transferred total leakage reactance on the secondary winding will be given as

$$X_s = x_p\prime + x_s = x_s + x_p\left(\frac{T_s}{T_p}\right)^2. \tag{4.7}$$

Leakage flux estimation is not easier due to the complexities of the leakage flux channel and the inability to make accurate measurements. To obtain an approximation equation that can be used, there are always assumptions that we need to make. The expression is validated and verified alongside the test data.

## 4.4 Evaluation of Leakage Reactance in a Core-Type Transformer Featuring Concentric Coils of the Same Length

The following assumptions are considered for the leakage reactance of the core-type transformer:

- Magnetizing current effect is ignored.
- The impact of iron saturation and reluctance is ignored.
- It is assumed that all the mmf is utilized to counteract the resistance posed by the coil height.
- It is presumed that the distribution of leakage flux in the coil and between low-voltage and high-voltage coils is along the leg axial path.
- Equal axial lengths for both primary and secondary windings.
- The flux paths are parallel to the winding along the axial height.
- $I_p\, T_p = I_s\, T_s =$ total mmf.
- The length of the mean turn of both windings is equal.

Let,

$b_p$ and $b_s$ = The radial depth of the primary and secondary coil.

$T_p$ and $T_s$ represent the number of turns/phase of the primary and secondary coils.

$I_p$ and $I_s$ represent the currents per phase on primary and secondary windings.

$L_{mtp}$ and $L_{mts}$ represent the mean length of turn of primary and secondary coils, respectively.

$L_{mt}$ represents the total mean length considering both the windings.

$L_0$ is the perimeter of the insulation for both the LV and HV coils.

$L_c$ is the axial tallness or span of both coils.

The total flux associated with either the LV or the HV coil arises from

- The leakage flux found within the low- and high-voltage coils.
- The leakage flux exists between the low- and high-voltage coils.

An elemental strip dx is examined at a distance "x" from the edge of the low-voltage winding to determine the flux linkage resulting from the flux within the coil.

Subsequently, the flux linkage of the low-voltage winding due to the flux $\varphi_x$ in the considered portion of the strip is analyzed.

$\Psi x = \varphi x$ multiplied by the number of turns associated with $\varphi_x$.

$\Psi$x = Ampere turns generating $\varphi$x multiplied by the permeance of the strip multiplied by the number of turns associated with $\varphi$x.

$$\Psi_{\mathrm{x}} = \left(\frac{I_{\mathrm{p}}T_{\mathrm{p}}x}{b_{\mathrm{p}}}\right) \times \left(\frac{L_{\mathrm{mtp}}dx\mu_0}{L_{\mathrm{c}}}\right) \times \frac{T_{\mathrm{p}}x}{b_{\mathrm{p}}}. \tag{4.8}$$

The average length of the strip is thought to be roughly equal to $L_{\mathrm{mtp}}$. Hence, the net resultant flux linkage caused by the overall flux generated within the winding is expressed as

$$\Psi_{\mathrm{x}} = \int_0^{b_{\mathrm{c}}}\left(\frac{I_{\mathrm{p}}T_{\mathrm{p}}x}{b_{\mathrm{p}}}\right) \times \left(\frac{L_{\mathrm{mtp}}dx\mu_0}{L_{\mathrm{c}}}\right) \times \frac{T_{\mathrm{p}}x}{b_{\mathrm{p}}}, \tag{4.9}$$

$$\Psi_{\mathrm{x}} = \frac{I_{\mathrm{p}}T_{\mathrm{p}}^2\mu_0 L_{\mathrm{mtp}}b_{\mathrm{p}}}{3L_{\mathrm{c}}}. \tag{4.10}$$

Assuming that half of the flux $\varphi_0$ connects to both the windings, the flux linking to the primary side of the winding due to half of $\varphi_0$ can be expressed as follows:

$$\begin{aligned}\Psi_{\mathrm{x}} &= 1/2\varphi_0 \times \text{Number of turns connected to } \varphi_0\\ &= \text{Ampere Turns generating } \varphi_0 \times \text{permeance of the pathway}\\ &\quad \times \text{number of turns associated with flux } \varphi_0\\ &= \frac{1}{2} \times I_{\mathrm{p}}T_{\mathrm{p}} \times \frac{L_0 a\mu_0}{L_{\mathrm{c}}} \times T_{\mathrm{p}}\end{aligned} \tag{4.11}$$

Hence, the net resultant flux linked to the considered primary side of the winding is,

$$\lambda = \psi + \psi_0 = I_{\mathrm{p}}T_{\mathrm{p}}^2 \times \frac{\mu_0}{L_{\mathrm{c}}}\left(\frac{L_{\mathrm{mtp}}b_{\mathrm{p}}}{3} + \frac{L_0 a}{2}\right). \tag{4.12}$$

Assuming that $L_{\mathrm{mtp}}$ and $L_0$ are equal, the total flux linked to the primary winding becomes with the assumption that $L_{\mathrm{mtp}} = L_0$.

$$\lambda = \psi + \psi_0 = I_p T_p^2 \times \frac{L_{\mathrm{mtp}}\mu_0}{L_c}\left(\frac{b_p}{3} + \frac{a}{2}\right). \tag{4.13}$$

Therefore, leakage reactance of the primary/phase,

$$x_{\mathrm{p}} = 2\pi f \times \left(\frac{\text{Flux Linkage}}{\text{Current}}\right)$$

$$x_{\mathrm{p}} = 2\pi f \times \left(\frac{I_{\mathrm{p}}T_{\mathrm{p}}^2 \times \frac{\mu_0}{L_{\mathrm{c}}}\left(\frac{L_{\mathrm{mtp}}b_{\mathrm{p}}}{3} + \frac{L_0 a}{2}\right)}{I_{\mathrm{p}}}\right)$$

$$x_{\mathrm{p}} = 2\pi f T_{\mathrm{p}}^2 \times \frac{\mu_0}{L_{\mathrm{c}}} \times \left( \frac{L_{\mathrm{mtp}} b_{\mathrm{p}}}{3} + \frac{L_0 a}{2} \right)$$

$$x_{\mathrm{p}} = 2\pi f \mu_0 T_{\mathrm{p}}^2 \times \frac{L_{\mathrm{mt}}}{L_c} \times \left( a + \frac{b_1 + b_2}{3} \right). \tag{4.14}$$

The leakage reactance of the secondary/phase,

$$x_{\mathrm{s}} = 2\pi f \mu_0 T_{\mathrm{s}}^2 \times \frac{L_{\mathrm{mt}}}{L_{\mathrm{c}}} \times \left( a + \frac{b_1 + b_2}{3} \right). \tag{4.15}$$

The total reactance is therefore given by

$$X_{\mathrm{p}} = x_{\mathrm{p}} + x_{\mathrm{s}}\prime = 2\pi f \mu_0 T_{\mathrm{p}}^2 \times \frac{L_{\mathrm{mt}}}{L_{\mathrm{c}}} \times \left( a + \frac{b_{\mathrm{p}} + b_{\mathrm{s}}}{3} \right). \tag{4.16}$$

## 4.5 Numerical with Solutions to Obtain Leakage Reactance and Percentage Regulation

***Example 4.1*** With the information below, calculate the leakage reactance of a 50 Hz transformer.

Looking at the primary turn, there are 510, and in the secondary turn, there are 30. Primary and secondary turn mean lengths are equal to 1.25 m and 1.05 m, correspondingly. The height of each winding is 0.61 m, width is 0.015 m, and thickness is 0.025 m.

**Solution**:

We know that the leakage reactance referred to primary side is given by

$$x_p = 2\pi f \mu_0 T_p^2 \times \frac{L_{mt}}{L_c} \times \left( a + \frac{b_1 + b_2}{3} \right)$$

From the given data,
$a = 0.015$, $L_{\mathrm{c}} = 0.61$ m
$b = 0.025$, $f = 50$ Hz
Hence,

$$L_{\mathrm{mt}} = \frac{1.25 + 1.05}{2} = 1.15 \text{ m}$$

$$x_{\mathrm{p}} = 2\pi(50)(4\pi \times 10^{-7})(510^2) \times \frac{1.15}{0.61} \times \left( 0.015 + 0.025 + \frac{0.025}{3} \right)$$

$$x_{\mathrm{p}} = 6.1\ \Omega$$

***Example 4.2*** A 600 kVA, 6600/400 V, 50 Hz, three-phase core-type transformer has the following data.

Width of LV and HV winding = 3 cm, width of duct between LV and HV winding = 2 cm, height of LV and HV winding = 40 cm, length of mean turn = 1.5 m, number of turns of HV winding = 220, calculate the leakage reactance of the transformer as it pertains to the high-voltage side.

**Solution**: Given

$$\begin{aligned} f &= 50\text{ Hz} \quad a = 0.02\text{ m} \\ T_1 &= 220 \qquad b_1 = 0.03\text{ m} \\ L_{\text{mt}} &= 1.5\text{ m } b_2 = 0.03\text{ m} \\ L_{\text{c}} &= 0.4\text{ m} \end{aligned}$$

$$x_{\text{p}} = 2\pi f \mu_0 T_{\text{p}}^2 \times \frac{L_{\text{mt}}}{L_{\text{c}}} \times \left(a + \frac{b_1 + b_2}{3}\right)$$

$$x_{\text{p}} = 2\pi (50)\left(4\pi \times 10^{-7}\right)(220^2) \times \frac{1.5}{0.4} \times \left(0.02 + \frac{0.03 + 0.03}{3}\right)$$

$$X_{\text{p}} = 2.866\Omega$$

***Example 4.3*** Determine the % regulation at full load and 0.8 pf lag for a 500 kVA, 6000/400 V Δ-Y, 3 φ, 50 Hz core-type transformer with equal-length cylindrical coils using the information below. Coil height is 48 cm, and their LV and HV coil thicknesses are 2.53 cm and 1.65 cm, respectively. There is 1.4 cm of insulation between the LV and HV coils. Coil mean diameter = 28 cm, voltage/turn = 8 V, and copper loss at full load = 3.8 kW.

**Solution**:

$$\begin{aligned} &\text{Number of primary turns, } T_1 = \frac{6000}{8} = 750 \\ &\text{Primary current, } \quad I_{\text{p}} = \frac{\text{kVA}\times 10^3}{3V_1} = \frac{500\times 10^3}{3\times 6000} = 27.7A \\ &\text{Primary resistance, } \quad R_{\text{p}} = \frac{\text{Copper loss}}{3I_{\text{p}}^2} = \frac{3.8\times 10^3}{3\times 27.7^3} = 165\ \Omega \end{aligned}$$

Reactance of primary winding,

$$x_{\text{p}} = 2\pi f \mu_0 T_{\text{p}}^2 \times \frac{L_{\text{mt}}}{L_{\text{c}}} \times \left(a + \frac{b_1 + b_2}{3}\right). \tag{4.17}$$

$$\begin{aligned} L_{\text{mt}} &= \text{Length of mean term of both primary and secondary winding} \\ &= \pi \times 28 \\ &= 87.96\text{ cm.} \end{aligned}$$

From Eq. (4.1)

$$x_{\mathrm{p}} = 2\pi(50)\left(4\pi \times 10^{-7}\right)(750^2) \times \frac{0.8796}{0.48} \times \left(0.014 + \frac{0.0165 + 0.0253}{3}\right)$$

$$x_{\mathrm{p}} = 11.35\ \Omega$$

And,

$$\%\text{Regulation} = \frac{I_{\mathrm{p}}R_{\mathrm{p}}\cos\emptyset + I_{\mathrm{p}}X_{\mathrm{p}}\sin\emptyset}{V_1} \times 100$$

$$\%\text{Regulation} = \frac{(27.7)(1.65)(0.8) + (27.7)(11.35)(0.6)}{6000} \times 100$$

$$\%\text{Regulation} = 3.75$$

## 4.6 Conclusion

Determination of resistance and leakage reactance of HV and LV windings is a key aspect in the design and performance efficiency of power transformers. These parameters give extra data on the significant transformer attributes, for instance, power losses, voltage regulation, short-circuit execution, and total reliability. In this paper, we considered the theoretical background, calculation methods to estimate these parameters, and practical analysis to calculate them as critical for transformer designs. The precise computation of resistance and leakage reactance in HV and LV windings and their consideration can be not only a technical necessity but also a foundation in transformer engineering. Improvement in analytical methodology and computational and experimental analyses remains constant, gradually uncovering how these parameters influence the transformers to ensure that new transformers are efficient, more reliable, and competent to meet the demands of power systems in the current society. The actualization of contemporary information technology and embracing sustainable practices are some of the challenges that the transformer industry can meet to develop a robust and green energy system. Thus, the results of this research emphasize the continued relevance of detailed design and computational strategies in defining the further evolution of power transformers.

## References

1. Nilesh Chothani DP, Raichura M (2023) Advancement in power transformer infrastructure and digital protection, 1st edn. Studies in Infrastructure and Control, Springer, Singapore
2. Chothani N, Raichura M, Patel D (2023) Transformer infrastructure for power grid. In: Advancement in power transformer infrastructure and digital protection. Springer Nature Singapore, Singapore, pp 1–26. https://doi.org/10.1007/978-981-99-3870-4_1

3. Patel D, Chothani N (2020) Digital protective schemes for power transformer. In: Power systems. Springer Singapore, Singapore. [Online]. Available: https://books.google.co.in/books?id=tjaSzQEACAAJ
4. Chothani N, Raichura M, Patel D (2023) An overview of the protection of power transformers. In: Advancement in power transformer infrastructure and digital protection. Springer Nature Singapore, Singapore, pp 27–69. https://doi.org/10.1007/978-981-99-3870-4_2
5. Kulkarni SV, Khaparde SA (2004) Transformer engineering: design and practice. Taylor & Francis
6. Flanagan WM (1993) Handbook of transformer design and applications. McGraw-Hill Education. [Online]. Available: https://books.google.co.in/books?id=--lSAAAAMAAJ
7. Standard I (2020) IEC 60076-1:2011 power transformers—part 1: general. International standard. 9782889124381
8. (2016) IEEE standard test code for liquid-immersed distribution, power, and regulating transformers. IEEE Std C57.12.90-2015 (Revision IEEE Std C57.12.90-2010), pp 1–120. https://doi.org/10.1109/IEEESTD.2016.7428800
9. (2006) A text book of electrical machines. Laxmi Publications. [Online]. Available: https://books.google.co.in/books?id=FLgMygrZDgEC
10. Chothani N, Raichura M, Patel D (2023) Total harmonic distortion-based improved transformer protective scheme. In: Advancement in power transformer infrastructure and digital protection. Springer Nature Singapore, Singapore, pp 159–185. https://doi.org/10.1007/978-981-99-3870-4_6
11. Raichura M, Chothani N, Patel D, Mistry K (2021) Total harmonic distortion (THD) based discrimination of normal, inrush and fault conditions in power transformer. Renew Energy Focus 36:43–55. https://doi.org/10.1016/j.ref.2020.12.001
12. Chothani N, Raichura M, Patel D (2023) Impact of transitory excessive fluxing condition on power transformer protection. In: Advancement in power transformer infrastructure and digital protection. Springer Nature Singapore, Singapore, pp 135–157. https://doi.org/10.1007/978-981-99-3870-4_5
13. Chothani N, Raichura M, Patel D (2023) Adaptive biased differential protection considering over-fluxing and CT saturation conditions. In: Advancement in power transformer infrastructure and digital protection. Springer Nature Singapore, Singapore, pp 187–229. https://doi.org/10.1007/978-981-99-3870-4_7
14. Vecchio RD, Vecchio RMD, Poulin B, Feghali PT, Shah DM, Ahuja R (2017) Transformer design principles, 3rd edn. CRC Press. [Online]. Available: https://books.google.co.in/books?id=4UIwDwAAQBAJ
15. Krishnan R (2001) Electric motor drives, modeling, analysis and control, 1st edn. Upper Seddle River, New Jersey, 07458: Prentice Hall, Inc
16. Silvester P, Konrad A (1973) Analysis of transformer leakage phenomena by high-order finite elements. IEEE Trans Power Appar Syst PAS-92(6):1843–1855. https://doi.org/10.1109/TPAS.1973.293564
17. Ohue M, Yasuo N, Takata M (2024) Innovations in mathematical modeling, AI, and optimization techniques. J Supercomput 81(1):340. https://doi.org/10.1007/s11227-024-06861-9
18. Chothani N, Patel DD, Desai A (2023) Hysteresis current control-based boost rectifier analysis for single phase solid state transformer. In: 2023 international conference on recent advances in electrical, electronics and digital healthcare technologies (REEDCON), pp 385–390. https://doi.org/10.1109/REEDCON57544.2023.10151262
19. Patel D, Desai A, Chothani N (2023) Fixed frequency current control-based boost rectifier analysis for single phase solid state transformer. In: 2023 international conference for advancement in technology (ICONAT), pp 1–6. https://doi.org/10.1109/ICONAT57137.2023.10080301

20. Patel DD, Chothani N, Parmar BD (2023) Design and development of solid state transformer with different current control techniques. In: 2023 IEEE 11th region 10 humanitarian technology conference (R10-HTC), pp 414–419. https://doi.org/10.1109/R10-HTC57504.2023.10461788
21. Karsai K (1987) Studies in large power transformers. Elsevier, Distribution for the U.S.A. and Canada, Amsterdam, New York, New York, NY, U.S.A. https://doi.org/10.1604/9780444995117

# Chapter 5
# Overall Design of Power Transformer

Design of power transformer is a systematic and detailed procedure aimed at the best electrical performance, efficient heat, and strong mechanical and economic. The design is based around critical components such as the core, windings, yoke, and structural frame. The starting point of the core design is to choose high permeability materials such as silicon steel to reduce hysteresis losses and increase magnetic flux. Structurally compact and efficient flux distribution is achieved by dimensional ratios such as core width-to-depth and window width-to-depth calculations. As the transformer window area is not fixed, the transformer window is required to accommodate high-voltage (HV) and low-voltage (LV) windings together with insulation and cooling provisions to minimize copper losses and effective heat dissipation. Calculation of the number of turns and conductor cross-sectional areas of the winding is done based on voltage, current, and the thermal limits is called winding design. Losses and voltage regulation are also evaluated to factors such as winding resistance, reactance, and leakage flux. The proportion of the yoke is configured to minimize flux leakage and balance the core. Next, all components are integrated into the transformer frame, taking into account space for insulating, cooling ducts, and structural integrity in order to absorb vibration and stress. At every step in the design, the durability, safety, and long-term, efficient operation are ensured by precise calculations and engineering judgment.

## 5.1 Introduction

Transformer design is more than electrical engineering, more of an electrical mechanic and thermal engineering activity, all works together to produce the best design. This section compiles relevant information of the basic concepts and methodologies relevant to power transformer design as well as the most current developments in the field.

 
N. Chothani et al., *Advancement in Infrastructure and Design Technology of Power Transformer*, Studies in Infrastructure and Control,
https://doi.org/10.1007/978-981-96-9420-4_5

The goals for building transformer design have a theoretical foundation that is supported by essential papers that focus on practical approaches. Lowdon gave detailed information on the methods that can be used in transformer design, recommendations advice on possible ways of improving core and winding [1]. These proposals outlined essential fundamentals of transformer core material with regard to its winding also describing procedural rudiments for later developments. These principles were further developed by Pansini with particular emphasis on the interaction of electrical transformers with the power system and the effects incurred by design decisions on the performance and useful life of equipment [2].

In the early years, Del Vecchio, Poulin, and colleagues entailed detailed consequences connected with core-form transformers, especially in the second edition of the book published in 2001, which paid more attention to efficiency and cost-effectiveness concerns [3]. Its subsequent editions (2010; 2018) incorporated newer developments in magnetic materials and computational methods, leading to much improved modeling and design of the transformer [3]. Dasgupta presented a comprehensive explanation of the transformer design with major emphasis on the construction of the winding that increased its efficiency and reduced the losses [4]. Building on this view, Kulkarni and Khaparde provided a more enhanced version of this perspective and related it with the current design practices and theory, presenting how thermal performance is linked with the electromagnetic characteristics [5]. Later, Heathcote's J & P Transformer book presents the most comprehensive information that an engineer may need regarding transformer operations and maintenance, as well as their construction [6]. Similarly, Sudhoff used multi-objective design approach to power magnetic devices to incorporate the optimization technology in achieving the trade-off solutions such as cost, efficiency, and thermal management [7].

For example, Zhu and Yang proposed a systematic investigation of advanced design techniques in transformers in 2021 to employ computing tools for better transformer core and winding design [8]. The authors Bell and Bodger presented a work in 2007 where they conducted performance metrics on magnetic circuit field using simulation-driven design methodologies by integrating magnetic circuit theory and finite element analysis [9]. In the most recent studies, more emphasis was placed on optimization of technology and fresh materials. Similarly, Hacan and Yildiz utilized finite element analysis (FEA) in the context of three-phase transformer design for increased efficiency via the utilization of iterative design [10]. Like Amoiralis [11] focused on loss, sound, and voltage regulation performance and implemented advanced materials and configurations to optimize the performance. For instance, Pramono and Wijaya sampled the transformer design problem and applied particle swarm optimization to its winding configurations, and achieved favorable results [12]. Adly and Hafiz have further applied this concept for multi-objective evolutionary optimization where Pareto optimality-based solutions are sought that optimize multiple conflicting criteria [13]. Smith and Allen noted some of the development made in yoke structure and magnetic field improvement, the availability of enhanced efficiencies industry [14]. Georgia Institute of Technology presented a new and innovative direction of transformer's design for the grid using various advanced materials along with the manufacturing methods [15].

Zhang and Wu [16] and Ebrahim et al. [17] included recent research on how hybrid and micro-transformer concepts hold promising for upcoming applications, including renewable energy systems. Cybersecurity and Infrastructure Security Agency (CISA) [18] and Electric Power Research Institute (EPRI) [19] have also pointed out that transformers will be even more important for achieving the required level of grid reliability under new operational conditions. The evolution of the design approach on transformers from basic methods to highly advanced optimization methods indicates the condition of the field. Thus, computational tools and techniques, advanced materials, as well as the development of new thinking approaches have continued to enhance the efficiency and reliability of power systems as a result of changing sentiments of the modern world.

## 5.2 Overall Design Process and Flowchart

Overall design of transformer includes core, winding, yoke, and also overall designing parameters calculations like distance between core centers, frame designing parameters. After complete calculation of the structure of the transformer, internal electrical parameters related like numbers of turns on primary and secondary winding (so called high-voltage and low-voltage windings), resistance, and reactance of the coils. Even the height of the low-voltage and high-voltage coils is also calculated.

As a starting of the procedure, provide the prelim data of the transformer designing requirement like kVA, frequency, numbers of phases, transformer type, core distances, types of core/yoke, flux, current density, etc. Generalized flow for design of transformer is elaborated in Fig. 5.1.

Some important notations and equations regarding to design of the transformer elaborated here as a sample equations,

First step is defined as the finding out the volts per turns means EMF/turn,

$$\frac{\text{EMF}}{\text{turn}} = K\sqrt{\text{kVA}}. \tag{5.1}$$

Here, $K$ is constant and it is defined as per $K = \sqrt{4.44fx \times 10^3}$.

Here value of $x$ also depends on the transformer type, construction, and service conditions.

Normalized value of $k$ is defined as per the types of transformer and it is given in the Table 2.1 (refer Chap. 2) or

$$E_\text{t} = \frac{1}{40}\sqrt{\frac{\text{kVA} \times 1000}{\text{Number of legs}}} \tag{5.2}$$

To find the net core area $E_\text{t} = 4.44fB_\text{m}A_\text{i}$.

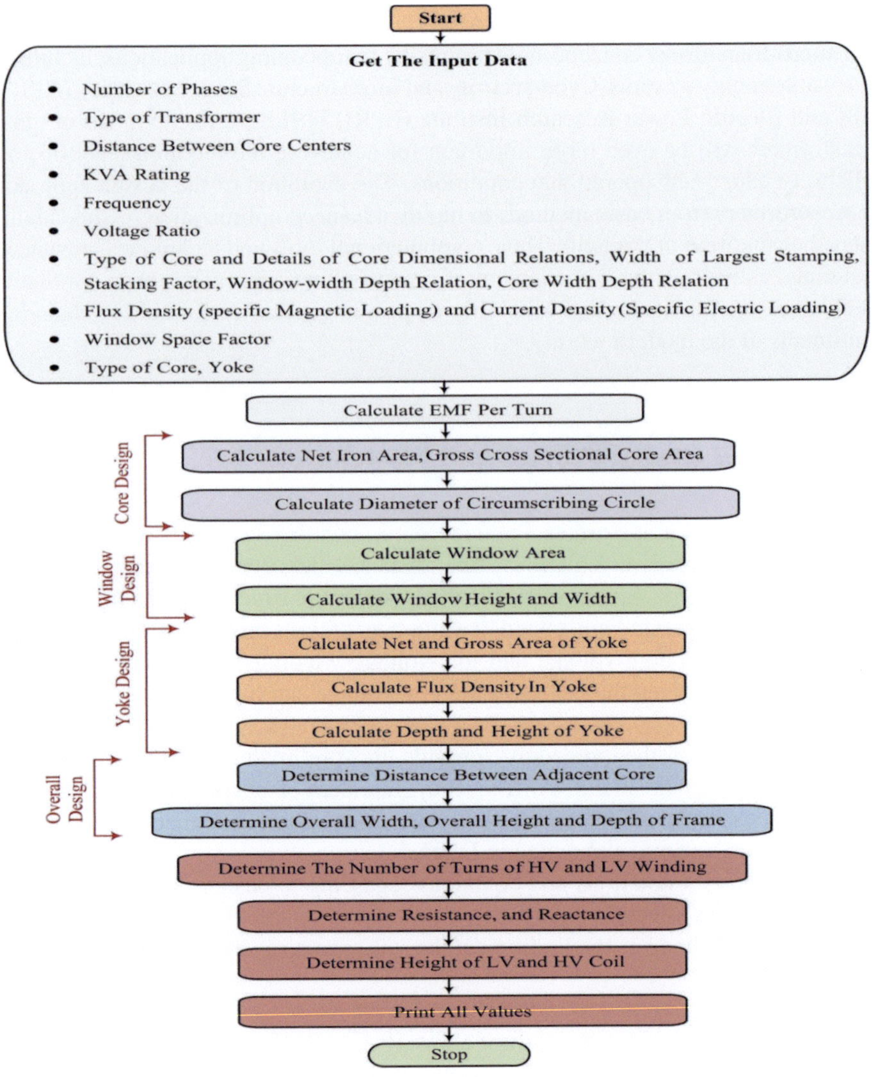


**Fig. 5.1** Flowchart for overall design procedure

$$A_i = \frac{E_t}{4.44 f B_m} \tag{5.3}$$

For gross cross-section area,

$$A_{gi} = \frac{A_i}{\text{Stacking factor}} \tag{5.4}$$

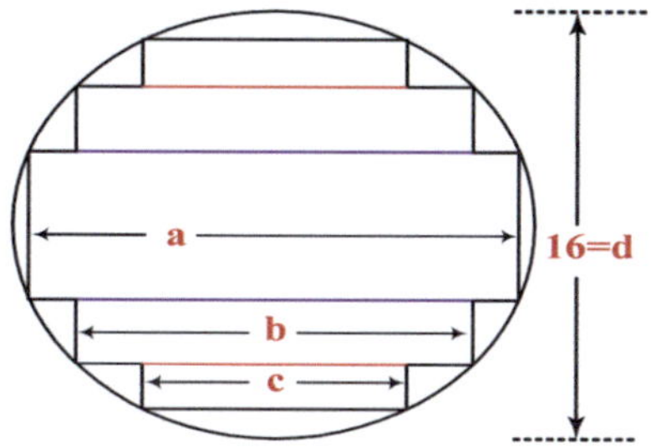


**Fig. 5.2** Three-stepped core

For three-stepped core as shown in Fig. 5.2, $A_i = 0.6d^2$.

Based on this, diameter of the circumscribing circle,

$$d = \sqrt{\frac{A_i}{0.6}} \tag{5.5}$$

Design of window:

$$Q = 3.33 f B_m A_i A_w k_w \delta \times 10^{-3} \tag{5.6}$$

$$\text{Window space factor} = \frac{\text{copper area in window}}{\text{Total window area}}. \tag{5.7}$$

For design of yoke, net iron area of yoke $= 1.2A_i$

$$\text{Height of yoke,}\ H_y = \frac{A_y}{W_y} \tag{5.8}$$

Overall dimension of core:

$$\text{Distance between adjacent core center} = D = W_w + d \tag{5.9}$$

$$\text{Overall width,}\ W = 2D + a \tag{5.10}$$

$$\text{Overall height,}\ H = H_w + 2H_y \tag{5.11}$$

$$\text{Number of turns,}\ T_1 = \frac{V_1}{V_2} \times T_2 \tag{5.12}$$

$$\text{Window space factor} = K_w = \frac{\text{Copper Area}}{\text{Window Area}} \tag{5.13}$$

## 5.3 Numerical Design Based on Given Data

Design a 100 kVA, 3$\varphi$, 50 Hz, 11 kV/440 V, Δ-Y distribution transformer, self-oil cooled with temperature rise not to exceed 40 °C. The no load loss is 350 W and copper loss and stay load loss on full load is 2000 W. The percentage impedance is 4.5%.

Also calculate, the no-load current, efficiency at 60 °C at full load, 75% of load at unity power factor and regulation at 75 °C at unity power factor.

**Solution**

**Volt/Turn**

$$\frac{\text{Emf}}{\text{turn}} = E_t = k\sqrt{\text{kVA}}$$

where $k = 0.45$ for 3$\varphi$ core-type distribution transformer

$$= 0.45\sqrt{100} = 4.5\,\text{V}$$

(or)

$$E_t = \frac{1}{40}\sqrt{\frac{\text{kVA} \times 1000}{\text{Number of legs}}} = \frac{1}{40}\sqrt{\frac{100 \times 1000}{3}} = 4.5\,\text{V/turn}$$

**Specific Magnetic Loading**

Let us choose $B_m = 1.35\,\text{Wb/m}^2$. Material for core can be hot-rolled steel laminations. Specific current loading:

Let current density $\delta$ be $\frac{2.5\,\text{A}}{\text{mm}^2} = 2.5 \times \frac{10^6\,\text{A}}{\text{m}^2}$.

**Cross-Section of Core**

$$E_t = 4.44 f B_m A_i$$

**Net Iron Area,**

$$A_i = \frac{E_t}{4.44 f B_m} = \frac{4.5}{4.44 \times 50 \times 1.35} = 0.015015\,\text{m}^2 = 150.15\,\text{cm}^2$$

**Gross Core Section Area,**

$$A_{gi} = \frac{A_i}{\text{Stacking factor}} = \frac{150.15}{0.9} = 166.83\,\text{m}^2$$

Let us choose a three-stepped core as shown in Fig. 5.2.

$$A_i = 0.6d^2$$

**Diameter of the Circumscribing Circle,**

$$d = \sqrt{\frac{A_i}{0.6}} = \sqrt{\frac{150.15}{0.6}} = 15.82\,\text{cm} \simeq 16\,\text{cm}$$

Let

$$\begin{aligned} d &= 16\,\text{cm} \\ a &= 0.9d = 14.4\,\text{cm} \\ b &= 0.7d = 11.2\,\text{cm} \\ c &= 0.42d = 6.72\,\text{cm} \end{aligned}$$

**Design of Window**

$$Q = 3.33fB_m A_i A_w k_w \delta \times 10^{-3}$$

$$\textbf{Window space factor} = \frac{\text{copper area in window}}{\text{Total window area}}$$

Let $k_w$ be 0.25.

**Window Area,**

$$\begin{aligned} A_w &= \frac{100}{(3.33)(50)(1.35)(150.15)(0.25)\left(2.5 \times 10^6\right) \times 10^{-3}} \\ &= 4.74 \times 10^{-6}\,\text{m}^2 = 474\,\text{cm}^2 \end{aligned}$$

Let the ratio of window height of window width be 2.5.

$$\begin{aligned} \frac{H_w}{W_w} &= 2.5 \\ A_w &= H_w W_w = 474 \\ (W_w)2.5W_w &= 474 \\ 2.5W_w^2 &= 474 \\ W_w &= 13.8\ \text{cm} \\ H_w &= 34.4\ \text{cm} \end{aligned}$$

The window design is shown in Fig. 5.3.

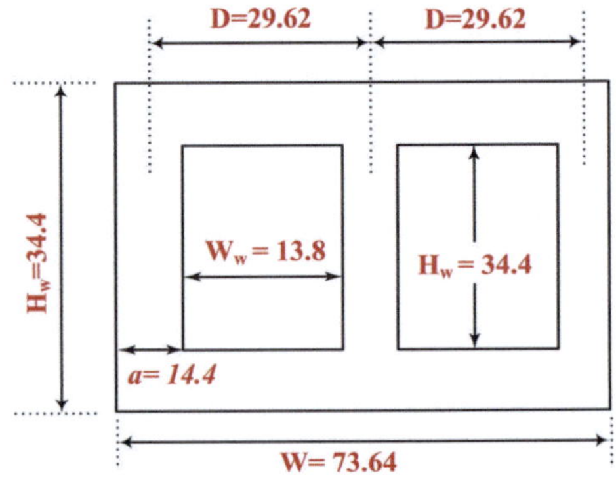


**Fig. 5.3** Window design with dimensions

**Design of Yoke**

$$\text{Net iron area of yoke} = 1.2A_\text{i} = (1.2)(150.15) = 180.18\,\text{cm}^2$$

$$\text{Gross yoke area} = \frac{180.18}{0.9} = 200.2\ \text{cm}^2$$

$$\text{Flux density in yoke} = \frac{1.35}{1.2} = 1.125\ \text{Wb/m}^2$$

Assuming yoke to be rectangular,

**Depth of Yoke,**

$$D_\text{y} = a = 14.4\ \text{cm}$$

**Height of Yoke,**

$$H_\text{y} = \frac{A_\text{y}}{W_\text{y}} = \frac{200.2}{14.4} = 13.9\,\text{cm} \simeq 14\,\text{cm}$$

**Overall Dimension of Core**

Distance between adjacent core center $= D = W_\text{w} + d$

$$= 13.8 + 15.82 = 29.62\,\text{cm}$$

**Overall Width,**

$$W = 2D + a = (2 \times 29.62) + 14.4 = 73.64\,\text{cm}$$

**Overall Height,**

$$H = H_w + 2H_y$$

**Depth of Frame,**

$$= 34.4 + (2 \times 14) = 62.4\,\text{cm}$$

## Design of Windings

### LV Winding

$$\text{Line voltage on LV side} = 440\,\text{V, Y connected}$$
$$\text{Phase voltage} = \frac{440}{\sqrt{3}}\,\text{V}$$
$$\text{Number of turns/phase} = \frac{254}{E_t} = \frac{254}{4.5} = 56.4 \simeq 57\,\text{turns}$$

Current per phase,

$$I_2 = \frac{100 \times 1000}{3 \times 254} = 131.23\,\text{A}$$

**Current Density,**

$$\delta = 2.5\,\text{A/mm}^2$$

$$\text{Area of conductors of LV winding} = \frac{I_2}{\delta} = \frac{131.23}{2.5} = 52.49\,\text{mm}^2$$

### HV Winding

$$\text{HV side line voltage} = 11{,}000\,\text{V}, \Delta\,\text{connected}$$
$$\text{Phase voltage} = 11{,}000\,\text{V}$$

**Number of Turns,**

$$T_1 = \frac{V_1}{V_2} \times T_2 = \frac{11{,}000}{254} \times 57 = 2468.5 \simeq 2469$$

$$\text{Current in HV winding} = \frac{100 \times 1000}{3 \times 11{,}000} = 3.03\,\text{A}$$

$$\text{Area of conductors of HV winding} = \frac{3.03}{2.5} = 1.212\ \text{mm}^2$$

$$\text{Diameter of conductor} = \sqrt{\frac{4}{\pi} \times 1.212} = 1.242\ \text{mm}$$

$$\text{Total copper area in window} = 2(a_1T_1 + a_2T_2)$$

$$= 2((52.49 \times 57) + (1.212 \times 2469))$$

**Window Space Factor,**

$$= 2(2991.93 + 2992.44) = 11,968.7\,\text{mm}^2 = 119.687\,\text{cm}^2$$

$$k_\text{w} = \frac{\text{copper area}}{\text{window area}} = \frac{119.68}{474} = 0.252$$

**Inside and Outside Diameter of LV Winding**

$$\text{Number of urns} = 57 \simeq 58$$

$$\text{Area of conductors} = 53.33\,\text{mm}^2$$

Let us choose copper rectangle conductors of size 6.3 × 4.5 mm (2 strips).

Let a paper insulation be used for these conductors. Let us take the inverse in dimension because of insulation, be 0.5 mm.

$$\begin{aligned}\text{Diameter of insulated conductor} &= (6.3 + 0.5) \times (4.5 + 0.5)\\ &= 6.8 \times 5 = 34\,\text{mm}\end{aligned}$$

Let us choose two layers.

$$\text{Turns per layer} = 29$$

Width of conductor is 6.8 mm along the height of the window $= 5 + 5 = 10\,\text{mm}$ along the width of the window.

For two layers, the conductor dimension is 6.8 mm along the height of window and with two conductors $5 + 5 = 10\,\text{mm}$ per paper width wise.

Helical winding is already used.

Therefore, space required is 28 + 1 times along the axial length.

$$\text{Axial length of winding} = 29 \times 6.8 = 197.2\,\text{mm} = 19.72\,\text{cm}$$

$$\text{Height of window} = 34.4\,\text{cm}$$

$$\text{Clearence} = \frac{34.4 - 19.72}{2} = 7.34\,\text{cm}$$

on each side of winding and will be filled with insulation.

Using 0.5 mm paper cylinders between layers,

$$\text{Radial depth of LV winding} = (2 \times 10) + 0.5 = 20.5\,\text{mm}$$

Diameter of circumscribing circle = 15.82 cm.

Using 1.5 mm thick, pressboard insulation between LV winding and core,

$$\text{Inside diameter of LV winding} = 15.82 + \left(2 \times 1.5 \times 10^{-1}\right) = 16.12\,\text{cm}$$

$$\text{Outside diameter} = 16.12 + (2 \times 2.05) = 20.22\,\text{cm}$$

$$\text{Mean diameter} = 18.17\,\text{cm.}$$

$$\text{Mean length of turn} = \pi \times 18.17 = 57.08\,\text{m.}$$

Inside and outside diameter of HV winding:

Maximum number of turns = 2468.

Size of conductor = 1.212 $\text{mm}^2$ round conductors,

With insulation, size of conductor = 1.375 (using enamel coating).

Let us choose number of coils for HV winding such that the volt/coil is nearly 1500 V.

If 8 disk coils are used,

$$\text{Volt/coil} = \frac{11{,}000}{8} = 1375.$$

$$\text{Turns/coil} = \frac{2468}{8} = 308.5 = 310\,\text{coils} \simeq 315$$

7 coils of 315 turns can be used and end coil of 263 turns can be used.

For normal coils, we can we 15 layers, turns per layer 21.

$$\text{Maximum voltage between layers} = 2 \times 21 \times 4.5 = 189\,\text{V}$$

(This is much below the normal voltage of 300 V between layers.)

$$\text{Axial depth of normal coil} = 21 \times 1.375 = 28.875\,\text{mm}$$

Axial depth of end coil = 15 layers, 18 turns per layer

$$= 18 \times 1.375 = 24.75\,\text{mm}$$

Let the spaces used between adjacent coils be 5 mm in height.

$$\begin{aligned}\text{Axial length of HV winding} &= (7 \times 28.875) + (1 \times 24.75) + (7 \times 5)\\ &= 261.875\,\text{mm}\end{aligned}$$

Height of window = 344 mm.
Clearance on either side = $\frac{344-261.875}{2}$ = 41.06 mm.
Insulation used between layers is 0.3 mm thick paper,

$$\text{Radial depth of HV coil} = (15 \times 1.375) + (14 \times 0.3) = 24.825\,\text{mm}$$

Total insulation between LV and HV winding = 15 mm.

(Bakelite paper = 5 mm, oil duct = 5 mm, kV insulation = 5 mm)

$$\text{Inside diameter of HV winding} = 202.5 + (2 \times 15) = 232.2\,\text{mm}$$

$$\text{Outside diameter of HV winding} = 232.2 + (2 \times 24.825) = 281.85\,\text{mm}$$

Clearance between two HV windings of adjacent lines = $D - 281.85$ = 14.35 mm. (This is sufficient for 11 kV.)

**Resistance**

$$\text{Mean diameter of HV winding} = \frac{232.2 + 281.85}{2} = 257.025\,\text{mm}$$

$$\text{Length of mean turn of HV winding} = \pi \times 257.025 = 807.47\,\text{mm} = 0.80747\,\text{m}$$

$$R_1 = \rho \frac{L_{m_1} T_1}{a_1} = \frac{0.021 \times .80747 \times 2468}{1.212} = 34.52\Omega$$

Mean diameter of LV winding = $\frac{161.2+202.2}{2}$ = 181.7 mm

$$\text{LV} = \pi \times 181.7 = 570.83\,\text{mm} = 0.57083\,\text{m}$$

$$\text{Length of mean turn of } R_2 = \frac{\rho L_{m_2} T_2}{a_2} = \frac{0.021 \times 0.57083 \times 56}{53}$$

$$= 0.0126\Omega.$$

Total resistance referred to primary side,

$$R_{ep} = R_1 + \frac{R_2}{k^2} = 34.52 + \frac{0.0126}{\left(\frac{56}{2468}\right)^2} = 58.9\Omega$$

$$\text{P.u resistance} = \frac{I_1 R_1}{V_1} = \frac{(3.03)(58.9)}{11{,}000} = 0.016$$

### Leakage Resistance

Mean diameter of windings $= \frac{181.7+257.025}{2} = 219.36\,\text{mm}$.

$$\text{Length of mean turn } = \pi \times 219.36 = 689.13\,\text{mm}.$$

### Height of HV Coil

Let us use a bakelite cylinder between LV winding and core of 5 mm thick. Therefore, the net distance between LV winding and core = 10 mm.

There are 8 coils on LV side with 15 layers and 21 turns/layer.

Height of HV winding over copper = (Height of HV winding in each disk × number of disk) + Height of spacer × number of spacer = (21 × diameter of conductor including insulation × number of disk) + Height of spacer × number of spacers: = (21 × 1.375 × 8) + (10 × 7) = 301 mm.

Height of LV coil: Number of turns of LV coils = 58.

This can be arranged in 2 layers with 29 turns in each layer.

The conductor size for LV is 6.8 × 10 mm.

$$\text{Height of LV over copper} = 29 \times 6.8 = 197.2\,\text{mm}$$

$$\text{Mean height of coil} = \frac{197.2 + 301}{2} = 249.1\,\text{mm}$$

$$\text{Ampere turns} = (131.23\,\text{A})(57\,\text{turns}) = 7480.11$$

$$\text{p.u reactance} = \frac{2\pi f \mu_0 L_{mt}(\text{AT})}{L_0 E_t}\left(a + \frac{b_1 + b_2}{3}\right)\text{p.u}$$

$$a = \text{distance between LV and HV coil} = 15\,\text{mm}$$

$$b_1 = \text{radial depth of LV winding} = 20.5\,\text{mm}$$

$$b_2 = \text{radial depth of HV winding} = 24.823\,\text{mm}$$

$$a + \frac{b_1 + b_2}{3} = 0.03108$$

$$= \frac{2\pi(50)\left(4\pi \times 10^{-7}\right)(689.13)(7480.11)(0.03108)}{(0.2491)(4.5)}$$

$$\text{p.u reactance} = 0.0564$$

$$\text{p.u impedence} = \sqrt{0.016^2 + 0.0564^2} = 0.05862\,\text{p.u}$$

## 5.4 Conclusion

The design of transformers in general is a careful integration of electrical, mechanical, and thermal concepts for effectiveness, durability, and economy. Every part and aspect of the transformer influences its performance, and various aspects are inextricably linked, demanding a design-systems approach. The transformer window design is therefore crucial in defining the probable magnetic flux path with space for the windings. Correct selection of the window area, width as well as depth allows for optimization of space and decreases leakage flux, and thereby prevents wastage of energy. The two established designs of cores and yoke are aimed at achieving high efficiency of magnetic utilization and using as lesser material as possible. The width-to-depth ratio and the stacking factor are among the most important dimensional relationships that have a critical impact directly on the functionality of the core to support the designed flux density without saturating or increasing excessive losses. The width of the largest stamping is well-controlled to strike a balance between manufacturability and structural requirements in the assembly. This is true because both HV and LV winding designs are equally important. Turns for the HV and LV windings are determined depending on the desired voltage transformation, insulation and spacing being important to avoid dielectric failure. Copper losses are controlled while voltage regulation is enhanced under different load conditions through proper choice of material and geometry for conducting the windings in a manner that offers the best of reactance. As for structural factors, the gross cross-sectional area of the core and net iron area of the core are determined to realize the needed flux density coupled with allowances for core losses. The geometric proportions of the window width to its depth and the core width to its depth help to improve the compact design yet be efficient. One factor, referred to as the stacking factor which defines the effective iron to total area ratio of core, measures material utilization against the deterioration of the magnetic performances. The total dimensions, that is, width, height, and depth of the transformer frame, depends on dimensions of the core and windings, provision for cooling, insulation, and mechanical strength. These dimensions must also be feasible in terms of transportation and installation, and conform to requirements of project delivery.

## References

1. Lowdon E (1988) Practical transformer design handbook. McGraw-Hill Companies [Online]. Available at: https://books.google.co.in/books?id=eLluAAAACAAJ
2. Pansini AJ (1999) Electrical transformers and power equipment. Fairmont Press [Online]. Available at: https://books.google.co.in/books?id=f77zWwA3oS4C
3. Vecchio RMD, Poulin B, Feghali PT, Shah DM, Ahuja R (2017) Transformer design principles: with applications to core-form power transformers, 2nd edn. CRC Press [Online]. Available at: https://books.google.co.in/books?id=45r0fFyK--kC
4. Dasgupta I (2002) Design of transformers. In: McGraw-Hill's Access Engineering. McGraw-Hill Education [Online]. Available at: https://books.google.co.in/books?id=8pxIzwEACAAJ

5. Kulkarni SV, Khaparde SA (2004) Transformer engineering: design and practice. Taylor & Francis
6. Heathcote M (2011) J&P transformer book. Newnes [Online]. Available at: https://books.google.co.in/books?id=paPKsOXn5FMC
7. Sudhoff SD (2021) Power magnetic devices: a multi-objective design approach. In: Power magnetic devices, 1st edn. Wiley. https://doi.org/10.1002/9781119674658.fmatter
8. Zhu B, Yang F (2021) Power transformer design practices, 1st edn. CRC Press, Tailor and Francis Group. https://doi.org/10.1201/9780367816865
9. Bell SC, Bodger PS (2007) Power transformer design using magnetic circuit theory and finite element analysis—a comparison of techniques. In: 2007 Australasian universities power engineering conference, pp 1–6. https://doi.org/10.1109/AUPEC.2007.4548092
10. Yildiz AB (2012) Electrical equivalent circuit based modeling and analysis of direct current motors. Int J Electr Power Energy Syst 43(1):1043–1047. https://doi.org/10.1016/J.IJEPES.2012.06.063
11. Amoiralis EI, Tsili MA, Georgilakis PS (2008) The state of the art in engineering methods for transformer design and optimization: a survey. J Optoelectron Adv Mater 10(5):1149–1158
12. Pramono WB, Wijaya FD, Hadi SP, Wahyudi MS, Indarto A (2023) Designing power transformer using particle swarm optimization with respect to transformer noise, weight, and losses. Designs 7(1). https://doi.org/10.3390/designs7010031
13. Adly AA, Abd-El-Hafiz SK (2015) A performance-oriented power transformer design methodology using multi-objective evolutionary optimization. J Adv Res 6(3):417–423. https://doi.org/10.1016/j.jare.2014.08.003
14. Smith TG, Allen JP (2023) Advancements in transformer yoke design and magnetic field optimization. J Electr Eng 15(3):123–134
15. Watson D (2023) ECE Team Honored for Groundbreaking Transformer Design. School of Electrical and Computer Engineering College of Engineering, Georgia Institute of Technology
16. Zhang LL, Wu QH, Ji TY, Zhang AQ (2017) Identification of inrush currents in power transformers based on higher-order statistics. Electr Power Syst Res 146:161–169. https://doi.org/10.1016/j.epsr.2017.01.029
17. Yang Z, Jia K, Fang Y, Zhu Z, Yang B, Bi T (2020) High-frequency fault component-based distance protection for large renewable power plants. IEEE Trans Power Electron 35(10):10352–10362. https://doi.org/10.1109/TPEL.2020.2978266
18. P. I. for the latest T. Updates (2024) Critical shortage of power transformers to ensure reliability of the U.S. Grid. 23. U.S. Cybersecurity and Infrastructure Security Agency (CISA)
19. Technical Update (2024) The copper book: a comprehensive guide for transformer operation, maintenance, and design, 1st edn. Electric Power Research Institute (EPRI)

# Chapter 6
# Effects of No-Load Current, Losses, and Frequency in Optimum Design of Transformer

Transformer design is vital in advancing electrical engineering in terms of performance improvement, reducing operational cost, as well as energy efficiency. This chapter offers a systematic approach to transformer design optimization based on accurate estimation of no-load current, core losses, core volume, and weight. The parameters are interrelated and have a great influence on transformer efficiency, cost of manufacturing, and durability. The relationships between key design variables such as core material characteristics, winding configurations, and geometric dimensions are established precisely through the combination of analytical and numerical methods. Core magnetization is analyzed in order to evaluate no-load current which accounts for hysteresis and eddy current losses. Standard performance criteria are used to minimize these losses to be realized by selecting core materials with high permeability and low loss factors. An optimization is made of the core volume and mass in order to achieve desired power ratings with reduced thermal and structural efficiencies. It is shown that this optimized design approach successfully reduces core losses and no-load current to lower the transformer's size and weight. Therefore, this methodology could support the development of high performance and economical transformers, the potential further improvement of which is enabled by the use of advanced materials and novel cooling technologies to make a more sustainable, more efficient power systems.

## 6.1 Introduction

Transformer design, especially in relation to the design of power transformers, focuses on the selection of core losses, no-load current, core weight, the effect of frequency variation, and types of material. It is critically important to better understand all these aspects and advance knowledge in the area in order to enhance transformer efficiency and reliability, and make the transformers able to fulfill the

 
N. Chothani et al., *Advancement in Infrastructure and Design Technology of Power Transformer*, Studies in Infrastructure and Control,
https://doi.org/10.1007/978-981-96-9420-4_6

continually increasing requirements of the electrical power systems. There exists a wealth of knowledge regarding these different facets in the literature and most of the discussions are centered on aspects such as advanced optimization procedures, design approaches, and materials.

Fundamental and practical methods for designing transformers have been developed, and these have been outlined in various books. Kulkarni and Khaparde in transformer engineering [1] provide a good reference to transformer design that covers issues like the choice of core material, computation of core loss, and ways for improving transformer efficiency. The authors of the paper also point out how avoiding core losses and optimizing the transformer windings is crucial for the improvement of the overall transformer performance. Say, in The Performance and Design of Alternating Current Machines, explains the transformer design in detail and employs the mathematical modeling approach in the analysis of the alternating current machines [2]. The book provides a relationship between the core losses, efficiency, and material properties of transformers, thus providing the ground for further research on the optimization of transformers. McLyman in the Transformer and Inductor Design Handbook took design principles a step further into the real world of operations of transformers [3]. He also stressed the need for concentration in the core areas and materials used, since it determines losses in the transformers used. McLyman's work depicts the growing complexity in the transformer design and the constant call for more sophistication to fit the current electrical standards. Vecchio, in Transformer Design Principles [4], aims to offer a discussion on core-form transformers and their subsequent optimization with context to core-form power transformers. He specializes in core-form transformers covering the effects of core construction and choice of materials on the transformer efficiency and performance.

The control of core losses is one of the critical determinants of the performance of transformers, which is a source of significant energy wastage. Shultz [5] scientifically explains core losses that are associated with the type of core material used and properties like permeability or saturation. Their work remains a basis for more complex modeling and design solutions that are intended to minimize these losses. Bhattacharyya [6] also includes the presentation of no-load current in the transformers and carries out an analysis of how the transformers are designed and how they affect the no-load conditions. The author describes the factors that influence no-load current and discusses the possibilities of transformer design to minimize it to save energy. Core loss reduction in transformers using advanced materials has considered core materials like amorphous steel and nanocrystalline alloys [7]. It has been seen that many of these materials can achieve a reduction in core losses as compared with conventional materials especially at the high-frequency range. The authors stress that material substitution plays a crucial role in the design of transformers.

Therefore, the reduction of transformer core's weight is crucial in achieving high efficiency in transformers without compromising the size or the expenses incurred by the transformers. Hassan and Keshmiri propose a methodology for the optimization of core weights that affect the efficiency of the transformers [8]. Thus, the authors present solutions that might help to decrease the level of transformer weight while simultaneously keeping its performance constant. Hussain and Gomez consider the balance that has to be struck between reducing the overall core weight and having enough core material to guarantee that the transformer will operate efficiently [9].

These studies show that variation affects an important aspect of the transformer, specifically core loss. In the influence of frequency on the core losses in electrical transformers, Liu, Zhao, and Li seek to understand how the frequency changes relate to core losses [10]. They deduce that with increased operating frequency, there is an increase in core losses, but this can be dealt with by proper material choice and reviewing the design. This factor is especially significant in present-day power systems where transformers are required to perform well at variable frequencies. Tang et al. [11] also studied the effects of temperature on the partial discharge behavior in transformer oil contaminated with metallic particles; it was found that increased temperatures facilitate discharge activity. This study furnishes useful suggestions for transformer insulation reliability under varied thermal conditions.

Advanced computational tools and optimization methods have brought drastic changes in designing the transformer. Amoiralis and Georgilakis in their article propose an ingenious method of using AI for transformer design [12]. Design parameters such as core material and winding configuration are determined by the use of the best AI algorithms that reduce losses and increase transformer efficiency. In modern parameter estimation techniques for power transformers, Gonzólez-Cagigal dedicate their work to another level of parameter estimation methods particularly on transformer loss estimation and minimization [13]. Their work shows how powerful engineering methods such as finite element analysis (FEA) and machine learning can be employed to accurately estimate transformer parameters to enhance design choices.

Reducing no-load losses is thus paramount to improving transformer efficiency. M. Foata discusses various techniques to minimize no-load losses in transformer efficiency enhancement by minimizing no-load losses [14]. The paper also illustrates the ways of enhancing the magnetic flux path as well as minimizing the core volume and shows that the problem is crucial for the future development of transformers. For the core type transformers designed in high frequency, Duppalli and Sudhoff present a fast and accurate way for computing the leakage inductance to improve the design efficiently. The proposed technique attains a favorable balance between the computational speed and the acceptable precision for practical engineering applications [15]. Based on finite element analysis, Kabas and Oguten optimize the design of three-phase transformer using detailed simulation-based evaluation, by which the performance and efficiency of the transformer are enhanced [16].

No-load current or magnetizing inrush current has direct effect on the design of protective schemes [17–19]. Different types of inrush, like sympathetic inrush, recovery inrush also give effect on the result of the differential protection [20–22]. However, discrimination for the inrush and internal fault is very clearly mentioned based on harmonic component [23, 24], but CT saturation conditions and over-fluxing conditions require more focus [25–27].

## 6.2 No-Load Current in the Transformer

The no-load current measurement is used to calculate the core losses of a transformer. This current, denoted as $I_0$, is illustrated as a phasor diagram in Fig. 6.1. The vector sum of the magnetizing current ($I_{\mathrm{m}}$) and the current component responsible for core losses or the working current ($I_C$) may be expressed as:

$$I_0 = \left(I_{\mathrm{m}}^2 + I_C^2\right)^{1/2}. \tag{6.1}$$

The magnetizing current, $I_{\mathrm{m}}$, generates the flux ($\varphi_{\mathrm{m}}$) in the magnetic circuit, while $I_C$ accounts for the no-load losses of the transformer. Hence, the no-load input power to the transformer is given by:

$$V_1(I_0 \cos \phi_0) = V_1 I_{\mathrm{C}}. \tag{6.2}$$

This is equivalent to the no-load loss since the output power is zero at the no-load condition. The general power relationship for the transformer is:

$$\text{Input} = \text{Output} + \text{Losses}.$$

Under no-load conditions, copper losses are minimal and can be considered negligible, meaning that the no-load losses are predominantly due to core losses. Therefore, the no-load current core loss component $I_{\mathrm{C}}$ is equal to the quotient of the core loss to the supplied voltage ($V_1$) since they apply to a single-phase transformer. For three-phase transformers, this component is calculated as the ratio of core loss per phase to the applied voltage ($V_{1\mathrm{ph}}$).

The root mean square (RMS) value of the magnetizing current, $I_{\mathrm{m}}$, is determined by the ratio of peak magnetizing ampere-turns to the $\sqrt{2}T_1$, with $T_1$ denoting the number of primary turns.

The iron path and air path are part of the magnetic circuit of the transformer. The iron path is from the legs and yokes, while the air path is due to the clearance created by joints in the core, which is made in stamped sheets of different shapes. Assuming all joints contribute equally to an air gap of length $l_{\mathrm{g}}$, the total ampere-turns for the transformer's magnetic circuit are given by:

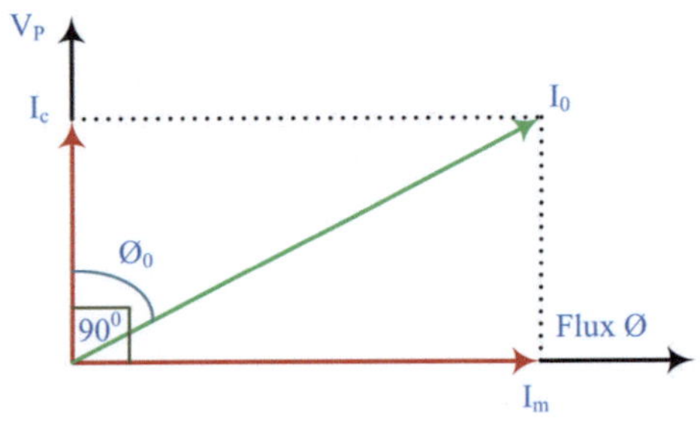


**Fig. 6.1** Phasor diagram for no-load current

$$\text{AT} = \text{AT(iron)} + 800,000\, l_g B_m$$

$$I_m = \frac{\text{AT for iron} + 800,000\ l_g\ B_m}{\sqrt{2}T_1}. \tag{6.3}$$

The transformer no-load current and its components in terms of mmf, number of turns, power, and voltage are derived for both single- and three-phase transformers as follows.

## 6.3 Magnetizing Current in Single-Phase Transformer

The magnetizing current's peak value in terms of mmf and number of turns is given by,

$$I_{m(pk)} = \frac{\text{AT}_0}{T_p}, \tag{6.4}$$

where

$\text{AT}_0$—Total magnetizing mmf under no-load condition, $T_p$—Number of turns in primary.

$$\begin{aligned}\text{AT}_0 &= \text{mmf of core} + \text{mmf of yoke} + \text{mmf of joints}\\ &= \text{AT}_c + \text{AT}_y + \text{AT}_j\end{aligned} \tag{6.5}$$

In the above equation,

$$\begin{aligned}(\text{AT}_c)\ \text{mmf of core} &= \text{mmf per metre for maximum flux density in core}\\ &\quad \times \text{ length of path traced by flux in core}\\ &= \text{at}_c \times 2l_c = 2\text{at}_c h_w\ \left[\because l_c = h_w(\text{height of window})\right]\end{aligned}$$

$$\begin{aligned}(\text{AT}_y)\ \text{mmf of yoke} &= \text{mmf per metre for maximum flux density in yoke}\\ &\quad \times \text{ length of the path traced by the flux in the yoke}\\ &= \text{at}_y \times 2l_y = 2\text{at}_y w_w\ \left[\because l_y = w_w\ (\text{width of window})\right]\end{aligned}$$

$\text{at}_c$, $\text{at}_y$—mmf per meter for maximum flux density in core and yoke, respectively

$2l_c$, $2l_y$—Total length of core and yoke, respectively

$$\text{AT}_j = 800,000 l_g B_m$$

(as all joints are presumed to be corresponding to an airgap ($l_g$)).

From Eq. (6.4), the rms value of magnetizing current for sinusoidal current is given by,

$$I_{\mathrm{m}}(\mathrm{rms}) = \frac{\mathrm{AT}_0}{\sqrt{2}T_{\mathrm{p}}}. \tag{6.6}$$

If the magnetizing current is non-sinusoidal, its rms values are given by,

$$I_{\mathrm{m}}(\mathrm{rms}) = \frac{\mathrm{AT}_0}{k_{\mathrm{peak}}\,T_{\mathrm{p}}}, \tag{6.7}$$

where $k_{\mathrm{peak}}$—peak factor.

### 6.3.1 *Core Loss Current*

The core or iron loss component of no-load current is given by

$$I_{\mathrm{c}}(I_{\mathrm{i}} \text{ or } I_{\mathrm{l}}) = \frac{P_{\mathrm{i}} \text{ or } P_{\mathrm{C}}}{V_{\mathrm{p}}},$$

where $P_{\mathrm{i}}(P_{\mathrm{c}})$—iron loss or core loss, $V_{\mathrm{p}}$—primary voltage.

## 6.4 Magnetizing Current in Three-Phase Transformer

The magnetizing current derived for a single-phase transformer holds good for a three-phase transformer also, but the length of the core is $3l_{\mathrm{c}}$.

### 6.4.1 *Magnetizing Volt-Ampere and Alternate Expression for Magnetizing Current*

The magnetizing volt-ampere $(\mathrm{VA})_{\mathrm{M}}$ is expressed as the product of voltage induced primary of transformer winding and magnetizing current.

Voltage induced in primary of transformer winding,

$$\begin{aligned} &= 4.44f\,\phi_{\mathrm{m}}T_{\mathrm{P}} \\ &= 4.44fB_{\mathrm{m}}A_{\mathrm{i}}T_{\mathrm{P}}\ [\because \phi_{\mathrm{m}} = B_{\mathrm{m}}A_{\mathrm{i}}] \\ \text{Magnetizing current} &= \frac{AT_0}{\sqrt{2}T_{\mathrm{P}}}. \end{aligned}$$

Hence, magnetizing volt ampere,

$$(VA)_M = 4.44fB_mA_iT_p \times \frac{AT_0}{\sqrt{2}T_P}. \tag{6.8}$$

Expressing mmf in terms of mmf per meter and length of iron path, we get,

$$\text{AT}_0 = \text{at}_i \times l_i. \tag{6.9}$$

Substituting Eq. (6.9) in Eq. (6.8), we get,

$$(\text{VA})_m = \frac{4.44fB_mA_i\text{at}_il_i}{\sqrt{2}}. \tag{6.10}$$

The magnetizing volt ampere per kg can be to found out by determining the ratio of magnetizing volt ampere to the weight of iron, which is given by,

$$W_i = \text{density} \times \text{volume}$$

We know that the density of iron is $7.8 \times 10^3$ kg/m$^3$ and volume is given by the multiplication of the area of iron ($A_i$) and the length of iron ($l_i$). Hence, using these values in the above equation, we get,

$$W_i = 7.8 \times 10^3 \times A_i \times l_i. \tag{6.11}$$

Hence, magnetizing volt ampere per kg, obtained by dividing Eq. 6.10 by Eq. (6.11), is

$$\begin{aligned}(\text{VA})_m/\text{kg} &= \frac{4.44fB_m\text{at}_i}{\sqrt{2} \times 7.8 \times 10^3}\\ &= 0.4025fB_m\text{at}_i \times 10^{-3}\\ &\simeq 0.4fB_m\text{at}_i \times 10^{-3}.\end{aligned}$$

Magnetizing current,

$$I_m = \frac{\text{Magnetizing VA/kg} \times \text{weight of core}}{\text{number of phases} \times \text{voltage/phase}}.$$

## 6.5 Numerical and Solutions

***Example 6.1*** With the help of the following data, find the no-load current of a 400 V, 50 Hz, $1\phi$ core-type transformer.

Length of mean magnetic path = 200 cm, gross core section = 100 $cm^2$, joints have 0.1 mm equivalent to air gap, maximum flux density = 0.7 Tesla, specific core loss = 0.5 W per kg at 50 Hz and 0.7 Tesla, ampere turns = 2.2 per cm, stacking factor = 0.9, density of steel = $7.5 \times 10^3$ kg/$m^3$.

**Solution**

We know that

$$
\begin{aligned}
\text{Total mmf} = \mathrm{AT}_0 &= 2\mathrm{at}_\mathrm{i} l_\mathrm{i} + 2\mathrm{at}_\mathrm{y} l_\mathrm{y} + \text{mmf for joints} \\
&= 2\mathrm{at}_\mathrm{i} l_\mathrm{i} + 2\mathrm{at}_\mathrm{y} l_\mathrm{y} + \frac{B}{\mu_0} l_\text{joints} \\
&= (\text{Amp. turns/m})\ (\text{length of mean magnetic path}) \\
&+ \frac{B}{\mu_0} 0.1 \times 10^{-3} \\
\mathrm{AT}_0 &= (2.2)(200) + (800,000)(0.7) \times 0.1 \times 10^{-3} \\
\mathrm{AT}_0 &= 496
\end{aligned}
$$

Net iron area,
EMF per phase,

$$
\begin{aligned}
A_\mathrm{i} &= \text{gross iron area} \times \text{stacking factor} \\
&= 100 \times 0.9 \times 10^{-4} \\
&= 90 \times 10^{-4}\,\mathrm{m}^2 \\
E_\mathrm{ph} &= 4.44 f B_\mathrm{m} A_\mathrm{i} T_\mathrm{ph} = 400\,\mathrm{V} \\
T_\mathrm{ph} &= \frac{400}{(4.44)(50)(0.7)\left(90 \times 10^{-4}\right)} \\
T_\mathrm{ph} &= 286 \\
I_\mathrm{m}(\mathrm{rms}) &= \frac{AT_0}{\sqrt{2} T_\mathrm{ph}} \\
&= \frac{496}{(\sqrt{2})(286)} = 1.226\,\mathrm{A}
\end{aligned}
$$

$$
\begin{aligned}
\text{Total weight of core} &= \text{volume of core} \times \text{density of steel} \\
&= \text{net core section} \times \text{mean length} \times \text{density of steel} \\
&= 90 \times 10^{-4} \times 2 \times 7.5 \times 10^3 = 135\,\mathrm{kg}
\end{aligned}
$$

Magnetizing current,

$$
\begin{aligned}
\text{Total core loss} &= \text{specific loss} \times \text{weight of core} \\
&= 0.5 \times 135 = 67.5\,\mathrm{W}
\end{aligned}
$$

Core loss current,

$$I_C = \frac{67.5}{400} = 0.168 \text{ A}$$

No-load current,

$$I_0 = \sqrt{I_m^2 + I_c^2} = \sqrt{1.226^2 + 0.168^2}$$
$$I_0 = 1.237 \text{ A}$$

***Example 6.2*** With the help of the following data, find the no-load current of a 200/100 V, 1 kVA, 50 Hz single-phase transformer. The gross cross-sectional area of core = 27 cm$^2$, effective magnetic core length = 0.5 m, weight of core = 7.5 kg, maximum flux density = 1.1 T, magnetizing mmf = 200 AT/m and specific core loss = 2 W/kg.

**Solution**

Rating, $Q = 1$ kVA, frequency, $f = 50$ Hz, gross core area, $A_{gi} = 27$ cm$^2$, core length = 0.5 m, weight of core = 7.5 kg, maximum flux density, $B_m = 1.1$ T, mmf = 200 AT/m and specific core loss = 2 W/kg.

We know that,

$$\text{Emf per phase} = 4.44 f B_m A_i T_{ph} = 200$$
$$\Rightarrow (4.44) f B_m (k_s A_{gi}) T_{ph} = 200$$

Hence,
Turns per phase,

$$T_{ph} = \frac{200}{(4.44)(50)(1.1)(0.9 \times 27 \times 10^{-4})} = \frac{200}{0.593} = 337$$

Also, we know that,

$$I_m = \frac{AT_0}{\sqrt{2} T_{ph}}$$

Let us assume stacking factor $k_i$ to be 0.9.

$$I_m = \frac{AT_0}{\sqrt{2} T_{ph}} = \frac{\text{magentizing mmf/m} \times \text{core length}}{\sqrt{2} T_{ph}}$$
$$= \frac{(200)(0.5)}{\sqrt{2} T_{ph}}$$

Substituting $T_{ph}$ in the above equation, we get,

$$I_m = \frac{100}{\sqrt{2} \times 337} = 0.209\,\text{A}$$

$$\text{Core loss current, } I_c = \frac{\text{core loss}}{v_1} = \frac{\text{Loss/kg} \times \text{weight of core}}{v_1}$$
$$= \frac{2 \times 7.5}{200} = 0.075\,\text{A}$$

$$\text{No-load current } I_0 = \sqrt{I_m^2 + I_c^2} I_0 = \sqrt{I_m^2 + I_c^2}$$

$$I_0 = \sqrt{0.209^2 + 0.075^2} = \sqrt{0.0436 + 5.625 \times 10^{-3}}$$
$$I_0 = 0.22186\,\text{A}.$$

***Example 6.3*** A 6600 V, 50 Hz, 1$\phi$ transformer has a sheet steel core. The cross-sectional area of iron is $\times\ 10^{-3}$ m$^2$. The maximum flux density is 1.2 T and the mean length of the iron path is 2.5 m. There are 4 joints in the core and each joint takes 40% as much reactive mmf, as it is required per meter of the core. Find the active, reactive components of no-load current, if magnetizing mmf/m is 232, specific core loss is 1.76 W/kg, and density is $5 \times 10^{-3}$ kg/m$^3$.

**Solution**

The magnetizing or reactive component of current,

$$I_m = \frac{\text{AT}_{\text{iron}} + \text{AT}_{\text{joints}}}{\sqrt{2} T_1} \tag{6.1}$$

$$\text{AT for core} = \text{mmf for core} \times \text{length of iron paths}$$
$$= 232 \times 2.5$$

AT for 1 joint = 40% AT/m for core,

$$\text{AT for 4 joints} = 4 \times \frac{40}{100} \times 232 \times 2.5$$

$$AT_{\text{joints}} = 928$$

$$\text{Number of primary turns } T_1 = \frac{V_1}{4.44 f \phi_m} = \frac{6600}{(4.44)(50)(24 \times 10^{-3})(1.2)}$$
$$(\because \phi_m = B_m A_i)$$
$$T_1 = \frac{6600}{6.393} = 1032$$

From Eq. (6.1), the magnetizing component of current is given by,

$$I_{\mathrm{m}} = \frac{(232)(2.5) + 928}{\left(\sqrt{2}\right)(1032)} = 1.0334\ \mathrm{A}$$

Active or core loss component of no-load current,

$$\begin{aligned}
I_{\mathrm{C}} &= \frac{\text{core loss}}{V_1} \\
&= \frac{\text{specific core loss} \times \text{volume of core} \times \text{density}}{6600} \\
&= \frac{1.76 \times 24 \times 10^{-3} \times 2.5 \times 7.5 \times 10^3}{6600} \\
&= 0.12\ \mathrm{A}.
\end{aligned}$$

No-load current,

$$I_0 = \sqrt{I_{\mathrm{m}}^2 + I_{\mathrm{c}}^2} = \sqrt{(1.033)^2 + 0.12^2}$$
$$I_0 = 1.0399\ \mathrm{A}$$

No-load line current $= \sqrt{3} \times 1.0399 = 1.801$ A.

## 6.6 Transformer Losses

The loss in the transformer has two components, namely load losses and no-load losses, as mentioned in the previous section.

### *6.6.1 No-Load Losses*

The no-load losses that occur in the core or iron part of the transformer occur at all times when the primary is energized, irrespective of whether the secondary is loaded or not. The $I^2R$ loss occurring on the primary due to no-load current is negligible. The core or iron loss, which is constant, has two components, namely hysteresis and eddy loss.

**Hysteresis loss**: This loss is due to magnetization and demagnetization of the iron core when it is subjected to an alternating magnetic field. This loss is directly proportional to the flux density $B_{\mathrm{m}}$, volume $V$ of the core and frequency $f$ of the alternating field.

Hysteresis loss,

$$P_{\mathrm{h}} = K_{\mathrm{h}} V B_{\mathrm{m}}^{1.6} f,$$

where $K_{\mathrm{h}}$ is a constant whose value depends on the type of magnetic material used.

**Eddy current loss**: Like hysteresis loss, the eddy current loss also occurs in the core of the transformer. The alternating magnetic flux produces an eddy current in the core, which results in a loss of power called eddy loss. To reduce this eddy current and hence the eddy loss, the core is laminated into thin sheets, which are insulated from each other by enamel coating. This loss is proportional to the volume of the core, the square of the flux density $B_{\mathrm{m}}$, frequency $f$ and also the square of the thickness of laminations, $t$.

$$\text{Eddy loss } P_{\mathrm{e}} = K_{\mathrm{e}} V B_{\mathrm{m}}^{2} f^{2} t^{2},$$

where $K_{\mathrm{e}}$ is a constant whose value depends on the resistivity of the core material used.

## 6.6.2 *Calculation of Core Loss*

Total core loss = core loss in legs + yoke's core loss

Estimation of the core losses can be done at the design stage by referring to the graph of core loss/kg versus flux density.

$$\begin{aligned}
\text{Core loss in leg } &= \text{loss/kg in leg} \times \text{weight of leg in kg} \\
&= \text{loss/kg in leg} \times \text{volume of the leg}(A_{\mathrm{i}} H_{\mathrm{w}}) \\
&\quad \times \text{density of steel or iron} \\
\text{Total core loss } &= \text{Number of legs} \times \text{loss/kg in leg} \times \text{volume of the} \\
&\quad \text{leg}(A_{\mathrm{i}} H_{\mathrm{w}}) \times \text{density of steel or iron}
\end{aligned}$$

Core loss in yoke = loss/kg in yoke × volume of yoke ($A_{\mathrm{y}}$ × mean length of the yoke) × density of iron used.

The iron or steel used for the transformer core has a density between 7.55 and 7.8 $\mathrm{g/cm^3}$.

## 6.6.3 *Calculation of Copper Loss*

Copper loss: These are the losses occurring in the two windings of the transformer under load conditions. Since these losses vary with the load, they are called load losses. Since this is the $I^2R$ loss in the windings, it can be calculated from the resistance and current rating of the winding.

### 6.6.4 Power Loss Formula

The power dissipated as heat due to resistance is given by Joule's law:

$$P = I^2R,$$

where
$P =$ Power loss (W),
$I =$ Current flowing through the conductor in amperes,
$R =$ Conductor's resistance (in ohms).

**Copper Loss in Transformer Windings**

The copper loss in the primary and secondary windings of the transformer are:

$$P_{\text{Primary}} = I_1^2 R_1$$

$$P_{\text{Secondary}} = I_2^2 R_2$$

where

- $I_1 =$ Current in the primary winding
- $R_1 =$ Resistance of the primary winding
- $I_2 =$ Current in the secondary winding
- $R_2 =$ Resistance of the secondary winding.

### 6.6.5 Total Copper Loss

The total copper loss in the transformer is the sum of the losses in the primary and secondary windings:

$$P_{\text{total}} = P_{\text{primary}} + P_{\text{secondary}}$$

$$P_{\text{Total}} = I_1^2 R_1 + I_2^2 R_2$$

## 6.7 Effects of Change in Frequency in Parameters of the Transformer

The input frequency of the transformer is one of the determining factors of the performance of the transformer. When the input frequency is changed, the transformer iron loss and hence the efficiency change. Similarly, the winding resistance, leakage reactance, and voltage are affected by the change in input frequency. The frequency change effects are discussed as follows.

1. **Effect on core loss**: Let us examine the effect of frequency variation on core loss if the voltage remains constant. We have the voltage, $E = 4.44fB_{\mathrm{m}}A_{\mathrm{i}}T$ and eddy current loss, $P_{\mathrm{e}} = K_{\mathrm{e}}f^2B_{\mathrm{m}}^2$.

If the voltage $E$ remains constant, then the product $(fB_{\mathrm{m}})$ remains constant. Thus, eddy current loss will remain constant as long as $E$ is kept constant, no matter the frequency is changed. But the hysteresis loss is proportional to the product $\left(fB_{\mathrm{m}}^x\right)$ and is given as,

**Hysteresis loss,**

$$P_{\mathrm{h}} = K_{\mathrm{h}}fB_{\mathrm{m}}^x.$$

The exponent $x$ can take the value from 1.6 to 2.2, depending on the magnetic characteristics of the core material.

If $x$ is assumed to be 2, then,

$$\begin{aligned}
P_{\mathrm{h}} &= K_{\mathrm{h}}(f\,B_{\mathrm{m}})B_{\mathrm{m}} \\
&= K_{\mathrm{h}}KB_{\mathrm{m}} \\
&\quad (\because E \text{ is constant}, f\,B_{\mathrm{m}} \text{ is constant and equal to } K) \\
&= K_{\mathrm{h}}K\frac{K}{f} = K_{\mathrm{h}}\frac{K^2}{f}.
\end{aligned}$$

Hence, hysteresis loss will decrease with an increase in frequency.

2. **Effect on winding resistance**: The winding resistance increases due to the skin effect when the frequency is increased. This increase is negligible when the change in frequency is very small.
3. **Effect on leakage reactance**: Since the leakage reactance $X_1 = 2\pi fl$, the leakage reactance changes linearly with the increase in frequency.
4. **Effect on voltage**: Along with the change in frequency, if the flux density is also changed, then a new voltage rating is possible. For the new voltage rating, the iron loss and no-load current can be calculated.

Table 6.1 gives a summary of the effects of frequency variation on various parameters discussed above.

**Table 6.1** Summary of effects of frequency variation on various parameters

| Parameter | Effect of change in frequency |
|---|---|
| Core loss | Decreases with an increase in frequency with voltage maintained constant |
| Voltage | Increases with an increase in frequency |
| Leakage reactance | Increases with an increase in frequency |
| Resistance | Increases with an increase in frequency |

## 6.8 Optimum Design of Transformer

A transformer design may be optimized to minimize overall volume, weight, cost and losses, etc. These requirements are generally conflicting, and so it is possible to satisfy any one of the above at one time.

We know that, the output equation of single-phase transformer is given by,

$$Q(\mathrm{kVA}) = 2.22 f B_{\mathrm{m}} \delta K_{\mathrm{w}} A_{\mathrm{w}} A_{\mathrm{i}} \times 10^{-3}$$
$$= 2.22 f B_{\mathrm{m}} \delta A_{\mathrm{c}} A_{\mathrm{i}} \; [\because K_{\mathrm{w}} A_{\mathrm{w}} = A_{\mathrm{c}}].$$

In the above equation, on assuming $f$, $B_{\mathrm{m}}$ and $\delta$ to be constant, the left out product of $A_{\mathrm{c}}A_{\mathrm{i}}$ is constant and is assumed to be given by,

$$A_{\mathrm{c}} A_{\mathrm{i}} = y^2. \tag{6.12}$$

We know that,

$$x = \frac{\phi_{\mathrm{m}}}{\mathrm{AT}}, \tag{6.13}$$

where

$$\phi_{\mathrm{m}} = B_{\mathrm{m}} A_{\mathrm{i}}. \tag{6.14}$$

And from Eq. (2.12),

$$\mathrm{AT} = \frac{\delta K_{\mathrm{w}} A_{\mathrm{w}}}{2} = \frac{\delta A_{\mathrm{c}}}{2}. \tag{6.15}$$

Substituting Eqs. (6.15) and (6.14) in Eq. (6.13), we get,

$$x = \frac{B_{\mathrm{m}} A_{\mathrm{i}}}{\frac{\delta A_{\mathrm{c}}}{2}} = \frac{2 B_{\mathrm{m}} A_{\mathrm{i}}}{\delta A_{\mathrm{c}}}.$$

On simplifying,

$$\Rightarrow \frac{A_{\mathrm{i}}}{A_{\mathrm{c}}} = \frac{\delta x}{2 B_{\mathrm{m}}} = z \text{ (Assumption)}. \tag{6.16}$$

The assumption is made such that $z$ is a function of $x$, as $B_m$ and $\delta$ are assumed to be constants. Using Eq. (3.53) in Eq. (3.49) to determine $A_i$ and $A_{c'}$, we get substituting $A_i = zA_c$ in Eq. (3.49), we get,

$$\begin{aligned} zA_c \times A_c &= y^2 \\ \Rightarrow \quad zA_c^2 &= y^2 \\ \Rightarrow \quad A_c &= \frac{y}{\sqrt{z}} \end{aligned} \tag{6.17}$$

Substituting $A_c = \frac{A_i}{z}$ in Eq. (6.12), we get,

$$\begin{aligned} \frac{A_i}{z} \times A_i &= y^2 \\ \Rightarrow \quad A_i^2 &= zy^2 \\ \Rightarrow \quad A_i &= y\sqrt{z} \end{aligned} \tag{6.18}$$

The total cost of transformer is given by,

$$C_T = c_i + C_c, \tag{6.19}$$

where, $C_i$—total cost of iron, $C_c$—total cost of copper conductor.
And

$$C_i = c_i g_i l_i A_i \tag{6.20}$$

$$C_c = c_c g_c l_{mt} A_c \tag{6.21}$$

where $C_i$—specific cost of iron, $C_c$—specific cost. Substituting Eqs. (6.20) and (6.21) in Eq. (6.19), we get,

$$C_T = c_i g_i l_i A_i + c_c g_c l_{mt} A_c. \tag{6.22}$$

Substituting Eqs. (6.17) and (6.18) in Eq. (6.22), we get,

$$C_T = c_i g_i l_i y\sqrt{z} + c_c g_c l_{mt} \tfrac{y}{\sqrt{z}}. \tag{6.23}$$

In order to determine the minimum cost, differentiating the above equation with respect to $z$, and equating it to zero, we get,

$$\begin{aligned} \frac{dC_T}{dz} &= \frac{1}{2} c_i g_i l_i \frac{y}{\sqrt{z}} - \frac{1}{2} c_c g_c l_{mt} \frac{y}{z^{-3/2}} = 0 \\ \Rightarrow \quad c_i g_i l_i \frac{y}{\sqrt{z}} &= c_c g_c l_{mt} \frac{y}{z^{3/2}} \end{aligned}$$

On simplifying the above equation, we get

$$c_i g_i l_i = \frac{c_c g_c l_{mt}}{z}.$$

Substituting Eq. (6.16) in the above equation, we get

$$\Rightarrow c_i g_i l_i = c_c g_c l_{mt} \frac{A_c}{A_i}.$$

$$\Rightarrow c_i g_i l_i A_i = c_c g_c l_{mt} A_c$$

$$c_i G_i = c_c G_c \tag{6.24}$$

Here, $G_i = g_i l_i A_i$, and $G_c = g_c l_{mt} A_c$

$$C_i = C_c. \tag{6.25}$$

Similarly, from Eqs. (6.24) and (6.25).
For the minimum volume of the transformer,

$$\text{Volume of iron} = \text{Volume of conductor}$$

$$\frac{G_i}{G_c} = \frac{c_c}{c_i}.$$

For the minimum weight of the transformer,

$$\text{Weight of iron} = \text{Weight of conductor}$$

$$G_i = G_c$$

For minimum losses of the transformer (maximum efficiency),

$$\text{Iron loss} = \text{Copper loss}$$

$$P_i = x^2 P_c$$

## 6.9 Conclusion

In the optimum design of transformers, the amount of no-load current, core loss and weight of the core, frequency variation, and other factors need to be determined. These factors are key determinants of the efficiency, performance, and cost of the transformer. Therefore, the no-load current that consists of magnetizing and core loss components must be correctly predicted in order to assess the performance of the transformer on a no-load circuit. The material properties and the frequency distribution selecting the core losses: energy losses, which take place in the core and which depend on material properties, frequencies, etc., and should be minimal for better efficiency, deciding at the time of selection of the material of the core. In the design of the transformer, the core weight is of utmost importance since it impacts the size and cost of the transformer. The light weight of the core can be achieved using lightweight materials and adopting an optimal design configuration that will minimize losses while ensuring sufficient structural integrity. Frequency fluctuation is also a key issue when it comes to the design of transformers, the core material of a transformer being frequency dependent and thereby leading to voltage drops and an increase in losses at different frequency levels from the standard rating. The transformer needs to accommodate these fluctuations in a way that its performance does not sharply degrade. The evaluation of these parameters in the design can be fine-tuned using the aid of computational models, artificial intelligence, and finite element analysis (FEA). This results in getting the most optimum, smallest, and cheapest transformer able to meet some specific performance characteristics. Moreover, meeting the standards and guidelines set in the industry proves that the transformer complies with safety, is reliable and is approved for use in the network system. In conclusion, the overall design of a transformer presents a number of correlated functions, including no-load current, core losses, core weight, frequency, and other parameters. Optimum design balances losses, lowers material expenses, and increases the operational life expectancy of the transformer. Integrating all the aspects of transformer designs guarantees that transformers adapt to the new standards of efficiency and endurance of modern electrical networks.

## References

1. Kulkarni SV, Khaparde SA (2004) Transformer engineering: design and practice. Taylor & Francis
2. Say MG (1949) The performance and design of alternating current machines, 2nd edn. Pitman & Sons
3. McLyman CWT (2011) Transformer and inductor design handbook. CRC Press, Boca Raton. https://doi.org/10.1201/b10865
4. Vecchio RMD, Poulin B, Feghali PT, Shah DM, Ahuja R (2010) Transformer design principles: with applications to core-form power transformers, 2nd edn. Taylor & Francis [Online]. Available at: https://books.google.co.in/books?id=YDGHZ_acR-EC

5. Shultz GP (1989) Introduction to transformers. In: Shultz GP (ed) Transformers and motors. Newnes, Boston, p 1. https://doi.org/10.1016/B978-0-08-051958-6.50020-2
6. Bhattacharyya M (2016) Electrical machines: modelling and analysis. Prentice Hall India Pvt., Limited [Online]. Available at: https://books.google.co.in/books?id=u3SADAAAQBAJ
7. Bolsi PC, Lenz JM, Prado EO, Piovesan T, Sartori HC, Pinheiro JR (2024) Cost and efficiency optimization of low-frequency transformers in uninterruptible power supplies. In: 2024 16th seminar on power electronics and control (SEPOC), pp 1–6. https://doi.org/10.1109/SEPOC63090.2024.10747421
8. Hassan MI, Keshmiri N, Callegaro AD, Cruz MF, Narimani M, Emadi A (2021) Design optimization methodology for planar transformers for more electric aircraft. IEEE Open J Ind Electron Soc 2:568–583. https://doi.org/10.1109/OJIES.2021.3124732
9. Hussain MK, Gómez P (2023) Optimal dielectric design of medium voltage toroidal transformer with electrostatic shield under fast front excitation. IEEE Trans Power Deliv 38(2):1395–1405. https://doi.org/10.1109/TPWRD.2022.3214101
10. Liu X, Zhao L, Ma C, Ge Q, Li Y (2022) Optimization simulation analysis of leakage magnetic field and loss characteristics of high frequency nanocrystalline transformer. In: 2022 25th international conference on electrical machines and systems (ICEMS), pp 1–6. https://doi.org/10.1109/ICEMS56177.2022.9983025
11. Pan C, Tang J, Zhang Y, Luo X, Li X (2018) Variation of discharge characteristics with temperature in moving transformer oil contaminated by metallic particles. IEEE Access 6:40050–40058. https://doi.org/10.1109/ACCESS.2018.2856580
12. Amoiralis EI, Georgilakis PS, Kefalas TD, Tsili MA, Kladas AG (2007) Artificial intelligence combined with hybrid FEM-BE techniques for global transformer optimization. IEEE Trans Magn 43(4):1633–1636. https://doi.org/10.1109/TMAG.2006.892258
13. Gonzólez-Cagigal MÁ, Rosendo-Macías JA, Gómez-Expósito A (2023) Parameter estimation for hot-spot thermal model of power transformers using unscented Kalman filters. J Mod Power Syst Clean Energy 11(2):634–642. https://doi.org/10.35833/MPCE.2022.000439
14. Foata M (2024) Economics of transformer and reactor life management. In: Cheim L, Gupta AK, MacArthur T-L, Ryder S (eds) Transformer and reactor life management. Springer, Cham, pp 79–114. https://doi.org/10.1007/978-3-031-77219-1_4
15. Duppalli VS, Sudhoff S (2017) Computationally efficient leakage inductance calculation for a high-frequency core-type transformer. In: 2017 IEEE electric ship technologies symposium (ESTS), pp 635–642. https://doi.org/10.1109/ESTS.2017.8069348
16. Hacan AF, Kabas B, Oguten S (2022) Design optimization of a three-phase transformer using finite element analysis [Online]. Available at: http://arxiv.org/abs/2201.11769
17. Patel D, Chothani N (2020) Digital protective schemes for power transformer, 1st edn. In: Power systems. Springer, Singapore [Online]. Available at: https://books.google.co.in/books?id=tjaSzQEACAAJ
18. Nilesh Chothani DP, Raichura M (2023) Advancement in power transformer infrastructure and digital protection, 1st edn. Studies in infrastructure and control. Springer, Singapore
19. Chothani N, Patel D, Raichura M (2019) Transformer protection with sequence components and digital filters, 1st edn. LAP LAMBERT Academic Publishing, Latvia, European Union
20. Chothani N, Raichura M, Patel D (2023) Introduction to magnetic inrush of power transformer. In: Advancement in power transformer infrastructure and digital protection. Springer, Singapore, pp 71–99. https://doi.org/10.1007/978-981-99-3870-4_3
21. Chothani N, Raichura M, Patel D (2023) An overview of the protection of power transformers. In: Advancement in power transformer infrastructure and digital protection. Springer, Singapore, pp 27–69. https://doi.org/10.1007/978-981-99-3870-4_2
22. Raichura M, Chothani N, Patel D (2021) Review of methodologies used for detection of magnetising inrush and fault conditions in power transformer. IET Energy Syst Integr 3:109–129. https://doi.org/10.1049/esi2.12012
23. Chothani N, Raichura M, Patel D (2023) Total harmonic distortion-based improved transformer protective scheme. In: Advancement in power transformer infrastructure and digital protection. Springer, Singapore, pp 159–185. https://doi.org/10.1007/978-981-99-3870-4_6

24. Raichura M, Chothani N, Patel D, Mistry K (2021) Total Harmonic Distortion (THD) based discrimination of normal, inrush and fault conditions in power transformer. Renew Energy Focus 36:43–55. https://doi.org/10.1016/j.ref.2020.12.001
25. Chothani N, Raichura M, Patel D (2023) Impact of transitory excessive fluxing condition on power transformer protection. In: Advancement in power transformer infrastructure and digital protection. Springer, Singapore, pp 135–157. https://doi.org/10.1007/978-981-99-3870-4_5
26. Chothani N, Patel Y, Patel J, Patel DD (2023) Adaptive numerical differential protection for transformer considering CT saturation and HIF. In: 2023 IEEE 3rd international conference on sustainable energy and future electric transportation (SEFET), pp 1–6. https://doi.org/10.1109/SeFeT57834.2023.10244915
27. Patel D, Chothani N (2020) CT saturation detection and compensation algorithm. In: Digital protective schemes for power transformer. Springer, Singapore, pp 33–49. https://doi.org/10.1007/978-981-15-6763-6_2

# Chapter 7
# Design of the Cooling System for Transformer

Thermal stability, operational reliability, and service life extension are dependent on effective cooling in the design of the transformer. Broadly, this chapter discusses different transformer cooling methods under the air-based and the oil-based cooling systems. The suitable methods for dry-type transformers and moderate heat dissipation are air natural (AN), air forced air natural (AFAN), and air blast (AB). For high-power transformers oil immersed cooling systems are offered by virtue of oil natural air natural (ONAN), oil natural air forced (ONAF) and oil natural water forced (ONWF), depending on the combination of natural convection and cooling media. High loading situations can be advanced cooled by oil forced air forced (OFAF), or oil forced air natural (OFAN) techniques for maximizing heat dissipation. The chapter also describes temperature rise in transformer tank with external cooling tubes, and reviews the thermal gradients and oil circulation. The material selection, geometry, and positioning as a means for optimizing the heat transfer in transformer tanks with tubes are covered. Practical numerical problems are also presented to show how to calculate temperature rise and how to determine the size of the cooling components, which reinforces the theoretical concept. The transformers are within safe temperature limits during different operating conditions with these methods.

## 7.1 Introduction

Hence, the fundamentals of power transformer design and operation remain crucial aspects of power system stability and performance. Ancillary equipment cooling systems in power transformers are essential in the operation and service life of the equipment by controlling the rate of losses by temperature rise during use. In the recent past, there has been a significant effort to try and optimize the thermal overlay as well as cooling methods to reduce effects like thermal degradation and increase the power levels. Several research articles and papers have been devoted to discuss

 
N. Chothani et al., *Advancement in Infrastructure and Design Technology of Power Transformer*, Studies in Infrastructure and Control,
https://doi.org/10.1007/978-981-96-9420-4_7

the cooling design and design improvement of power transformers considering both experimental and computational approaches. Transformer infrastructure regarding to protective points of view elaborated in depth with different protective schemes [1–4].

In the work [5], Zhu and Yang focus on practical aspects of the transformers with focus on the recent cooling technologies to solve the thermal issues. Kulkarni and Khaparde [6] explain the diagnostic and cooling systems in engineering transformers, the connection between conventional ways, and advanced technologies. Likewise, Harlow's edited volume [7] presents information on fundamental aspects of transformer engineering, such as thermal management strategies are documented. Del Vecchio et al. [8] focus on the core-form transformer design, including the cooling option to enhance transactional use. In his outstanding review of applications of transformers [9], Winders emphasizes the cooling issue as vital in handling thermal loads. One common approach in cooling is described as the central theme in managing temperature rise; Flanagan [10] on practical transformer design. Moreover, Lopez-Fernandez [11] and McLyman [12] explore the theoretical and numerical aspects in the thermal characterization of transformers while offering tools to accurately model heat generation. Dasgupta [13] elaborates on quality assurance activities that include cooling system analysis. As highlighted in the Springer Nature publication [14], improvements in the transformer model in recent years incorporate contemporary cooling methods to optimize it.

With regards to the concepts of transformer cooling, the research articles contribute to the enhancement of knowledge in this same. Khan et al. [15] designed an adaptive cooling system that can automatically adapt to the changes in operation. According to Goscinski et al. [16], there are important design aspects in cooling systems that should be considered in relation to the performance of the transformer. Ngo et al. [17] present a forced air-cooling approach for high-power transformers with a discussion on scalability. Literature reviews, such as Saket et al. [18] and Prajapati et al. [19], support the argument about the necessity of cooling for increasing the lifetime of transformers. Kim et al. [20] use computational simulations in order to enhance cooling designs. Sacco et al. [21] put forward a multiscale model that will enhance the conditions of two-phase cooling. Even during real-time monitoring of the transformer, temperature is the base value to analyze the system [22–24]. Even temperature effect also affects the behavior of solid-state transformer as it solely works with power electronics parts [25–27]. These studies together show an evolution of the research on transformer cooling from the experimental approach to the highly complex numerical simulations. The introduction of smart systems and the use of optimization techniques demonstrate the continuous development of transformers that will continue to support the grid in providing the necessary power capacity needed in the face of ever increasing operation.

## 7.2 Types of Cooling in Transformers

The transformer components heat up as a result of losses that form in the winding and core. Depending on the method of cooling, the heat is transferred to the cooling medium according to the thermal gradient. In a very small transformer of a few kVA rating, the cooling surface of the core and winding is sufficient to be cooled by normal circulation of atmospheric air. A more effective cooling system is necessary since losses rise in conjunction with the transformer's capacity. Figure 7.1 displays flowchart illustrating the types of cooling used in electrical transformers, principally differentiating between dry type and oil filled transformers. Moreover, Fig. 7.2 shows various diagrams related to transformer tanks and their cooling mechanisms mainly focusing on oil circulation and radiator.

### Air Natural (AN)

- Relative to the other methods, this one make use of air as the cooling agent.
- Natural convection is used in the circulation of the surrounding air for the removal of heat developed by natural circulation.
- A sheet metal enclosure is fitted to shield the windings from mechanical damage affecting it.
- This technique is applicable for small low-voltage transformers with ratings of up to 1.5 MVA.

### Air Forced Air Natural (AFAN)

- This technique uses a fan to move air through the transformer, and natural air circulation cools the air in a heat exchanger.
- This method is not commonly used.

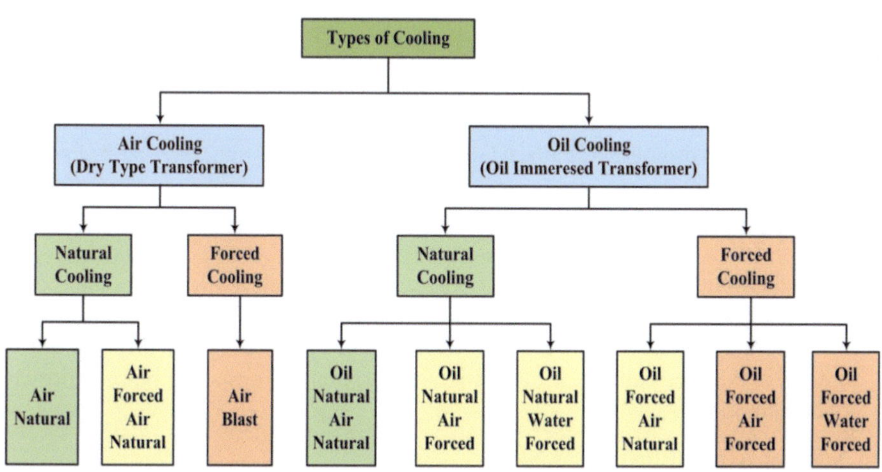


**Fig. 7.1** Types of cooling system

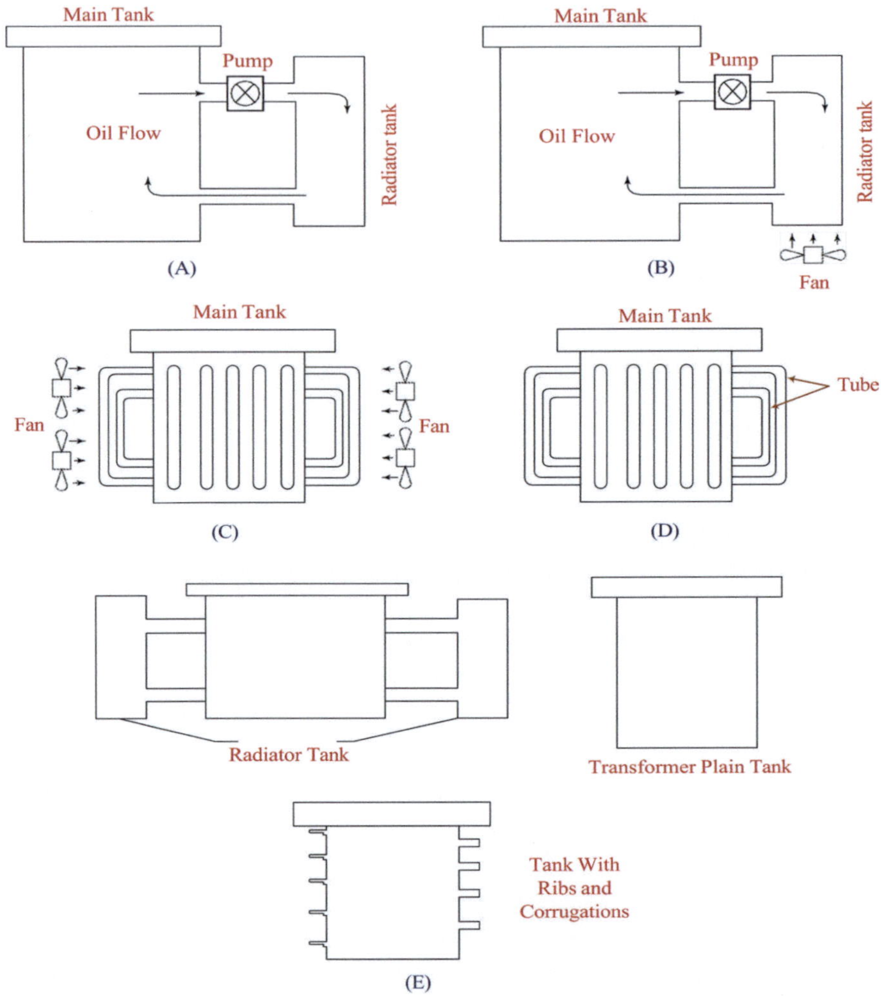


**Fig. 7.2** Cooling mechanisms: **a** OFAN, **b** OFAF, **c** ONAF, **d** ONAN with tubes, and **e** ONAN

### Air Blast (AB)

- In this technique, cool air is blown continuously through the windings and cores in order to cool the transformer.
- External fans cause the blowing of the air bursts. Specifically, the air-type transformers make it possible to achieve higher specific loadings as air blast improves heat dissipation.

### Oil Natural Air Natural (ONAN)

- Oil is a better conductor of heat than air.

- The coefficient of volume expansion of oil with respect to temperature is quite significant.
- Oil and air are naturally circulated in this cooling.

**Oil Natural Air Forced (ONAF)**

- In this technique, the heat is conducted through oil circulation and reaches the tank walls.
- In order to cool the transformer, forced air is circulated through the space vacated from the transformer tank.

However, the normal way of cooling the transformers by air blast is by providing radiation tanks which are either of corrugated or elliptical tubes in separate from the transformer tank, which is cooled by the air blast produced by fans.

**Oil Natural Water Forced (ONWF)**

- This technique requires the placement of copper cooling coils below the oil's surface and above the transformer.
- The transformer is then cooled by circulating water through with the help of the cooling coils.
- However, where a natural water head is already available, this measure proves more economical.

**Oil Forced Air Natural (OFAN)**

- For large transformers, forced circulation is used because natural oil circulation is inadequate to cool these transformers.
- From the top of a transformer tank, oil is circulated by a motor-driven pump to an external cooling plant for cooling.

**Oil Forced Air forced (OFAF)**

- Heat is transferred to the tank walls by the warm oil circulating in it, under natural heat, in the oil natural–air forced method.
- The air is then forced through the void in order to cool the transformer.
- The heated oil is cooled by an air blast furnished by fans in an external heat exchanger.
- Therefore, mixed cooling conditions are used for the transformer working with oil-natural air-natural conditions up to 50% rating and forced air forced oil conditions at higher loads.

**Oil Forced Water Forced (OFWF)**

- By so doing, the oil pressure is effectively kept higher than the water pressure as a technique.
- This cooling method proves convenient for use in the transformer banks; however, such a system's reliability prevents the connection of more than, for instance, three tanks to one cooling pump circuit.
- OFWF is used for transformers intended for use in a hydroelectric power plant.

**Table 7.1** Types of heat transfer from various regions of transformer

| From region | To region | Type of heat transfer |
|---|---|---|
| Core and winding | Outer surface, yoke | Conduction |
| Outer surface, yoke | Oil | Convection |
| Oil | Tank or cooler walls | Convection |
| Tank or cooler walls | Cooling medium (air or water) | Convection and radiation |

## 7.3 Temperature Rise and Design of the Cooling System

Tanks with plain walls are used for smaller rating transformers up to 20–30 kVA rating, as they have sufficient surface area to dissipate the heat. However, in the case of the basic plain-walled tanks, the transformers have a higher rating, and that the width increase in temperature will be above the acceptable working limit. Hence, tubes and radiators are used. The tank walls, tubes, and radiators dissipate heat by radiation and convection. The modes of heat transfer from various regions of transformer are given in Table 7.1.

## 7.4 Plain-Walled Tank's Temperature Rise

Some of the recommended maximum operating temperatures include, for natural cooling, the maximum rise should not be more than 55 °C, for forced air 60 °C, forced water 65 °C, while for oil cooling is at 45 °C. This means per $m^2$ of tank surface, the cooling system wastes 6.5 W by convection and 6 W by radiation per degree Celsius. Generally, for heat dissipation, the bottom surface is not considered. Similarly, the top surface has fittings, bushings, etc., and is not considered to dissipate heat. Hence, the cooling capacity of 12.5 St watts per $m^2$ per degree Celsius will be dissipated if the transformer tank cooling surface area is considered to be St $m^2$, excluding the top and bottom surface.

Hence, for a plain-walled tank, the temperature rise,

$$\theta = \frac{\text{Total loss in Watts}}{12.5\ S_t}. \tag{7.1}$$

## 7.5 Temperature Rise in the Tank with Tubes

Losses and temperature rises are proportional to the transformer's capacity. It could be necessary to blow air over the transformer to limit the temperature increase. This is not recommended because the insulation may be impacted by the ambient air that contains moisture, oil particles, etc. The transformer is positioned inside an oil-filled steel tank to mitigate the issue of atmospheric dangers. The oil takes heat from the coil and core and passes it to the wall of the tank. Heating is removed from the walls of the tank in the surrounding environment by radiation and convection. Additionally, the transformer dissipating area's capacity can be extended by utilizing tubes or auxiliary radiator tanks without expanding the tank's size.

Let the cooling surface of the tank be $S_t m^2$. This will dissipate 12.5 $S_t W/m^2/°C$. By provision of tubes, the total surface area becomes $xS_t m^2$.

Convection from tube walls $= xS_t 6.5 \times 1.35$ W $= xS_t 8.8$ W($\because$ 1.35 is due to a 35% increase in heat transfer due to convection by syphoning action of the tubes)

$$\text{Total dissipation} = 12.5S_t + xS_t 8.8 \text{ W}. \tag{7.2}$$

The number of tubes can be determined by knowing the surface area of each tube. The diameter of tubes normally used is 5 cm and they are spaced 7 to 7.5 cm center to center apart.

## 7.6 Design of Tank with Tubes

$$\text{Tank's dissipating surfacc area} = S_t$$

$$\text{Tank's dissipating surface area} = xS_t$$

$$\text{Total loss dissipation at the surface area of the tank} = 12.5\ S_t W/°C$$

$$\text{Total loss dissipation in surface area of and cooling tubes} = \frac{(12.5 + 8.8x)S_t \text{ W}}{°C}$$

$$\begin{aligned}\text{Total loss dissipation by tubes due toconvection} &= 1.35 \times 6.5xS_t \\ &= 8.8xS_t \text{ W}/°C\end{aligned}$$

$$\begin{aligned}\text{Tank and tube walls total surface area} &= xS_t + S_t \\ &= (1 + x)S_t\end{aligned}$$

$$\begin{aligned}\text{Total loss dissipation} &= \frac{\begin{array}{c}\text{Total loss dissipation by in surface}\\ \text{area of tank and cooling tubes}\end{array}}{\text{Total surface area of walls of tank and tubes}} \\ &= \frac{(12.5 + 8.8x)S_t}{(1 + x)S_t}\end{aligned}$$

Temperature rice with tubes,

$$
\begin{aligned}
(\theta) &= \frac{\text{Total loss}}{\text{Specific heat dissipation } \times \text{surface area}} \\
&= \frac{P_{\mathrm{i}} + P_{\mathrm{c}}}{(12.5 + 8.8x)S_{\mathrm{t}}}
\end{aligned}
$$

From the above equation, area of tubes can be found as follows.

$$
\begin{aligned}
\theta &= \frac{P_{\mathrm{i}} + P_{\mathrm{C}}}{12.5S_{\mathrm{t}} + 8.8xS_{\mathrm{t}}} \\
&\Rightarrow 12.5S_{\mathrm{t}} + 8.8xS_{\mathrm{t}} = \frac{P_{\mathrm{i}} + P_{\mathrm{c}}}{\theta} \\
&\Rightarrow 8.8xS_{\mathrm{t}} = \frac{P_{\mathrm{i}} + P_{\mathrm{C}}}{\theta} - 12.5S_{\mathrm{t}}
\end{aligned}
$$

Tube's total area,

$$
xS_{\mathrm{t}} = \frac{1}{8.8}\left[\frac{P_{\mathrm{i}} + P_{\mathrm{c}}}{\theta} - 12.5S_{\mathrm{t}}\right] \tag{7.3}
$$

Length of a tube $= l_t$

Diameter of a tube $= d_t$

Radius of a tube $= r_{\mathrm{t}}$

Area of a tube $= \pi d_{\mathrm{t}} l_{\mathrm{t}}$

Total number of tubes,

$$
\begin{aligned}
n_{\mathrm{t}} &= \frac{\text{Tube's total area}}{\text{Tube's area}} \\
&= \frac{1}{8.8\pi d_{\mathrm{t}} l_{\mathrm{t}}}\left[\frac{P_{\mathrm{i}} + P_{\mathrm{C}}}{\theta} - 12.5S_{\mathrm{t}}\right].
\end{aligned} \tag{7.4}
$$

The voltage of the transformer and its rating, the electrical distance between the transformer and the tank, the distance of the base and oil level above the transformer, and the distance available for various connections and taps are the factors that determine the overall size of the tank. The clearance between different parts depends on the rating of the transformer.

Let,

$$D = \text{distance between adjacent limbs}$$

$$D_{\mathrm{e}} = \text{external diameter of outer winding}$$

**Table 7.2** Clearance values for a typical transformer

| Clearances | $C_w$ | $C_l$ | $C_h$ |
|---|---|---|---|
| Values ( mm) | 40 to 80 | 50 to 125 | 400 to 600 |

$C_w$ = the distance clearance between the outer winding and the wall of the tank concerning their width

$C_l$ = Lengthwise clearance of the space available between the tank wall and the outer winding

$C_h$ = Height - wise clearance between the tank wall and the outer winding.

The various clearance values for a typical transformer of rating 1000 to 5000 kVA and 11 to 33 kV voltage rating is given in Table 7.2.

The height-wise clearance includes the clearance of 50 to 60 mm at the base, clearance of 150 to 250 mm above the oil and space of about 200 to 250 mm for leads. Figure 7.3 shows the transformer main dimensions with the tank along with various clearances.

## 7.7 Flowchart for Design of Cooling System

The procedure for the design of the cooling system is given below as a flowchart in Fig. 7.4. This involves the calculation of the number of cooling tubes required and suitably arranging them around the tank.

## 7.8 Numerical on Design of Transformer Cooling

***Example 7.1*** Design a cooling system for a 3ϕ oil-immersed transformer with a delta/star core-type rating of 200 kVA, 6000/400 V, 50 Hz. For the tank walls, a temperature rise of 50 °C is allowable and the limits of maximum stress are determined. The tank has a height of 125 cm and a width of 100 cm, 50 cm in length. However up to five kW of losses are allowed. If cooling system is not present, to what degree the temperature can increase?

**Solution**:

Given, transformer rating, $Q$ and type = 200 kVA and $3\phi$

Voltage rating = 6600/400 V

Acceptable temperature rise = 50 °C

Acceptable losses = 5 kW

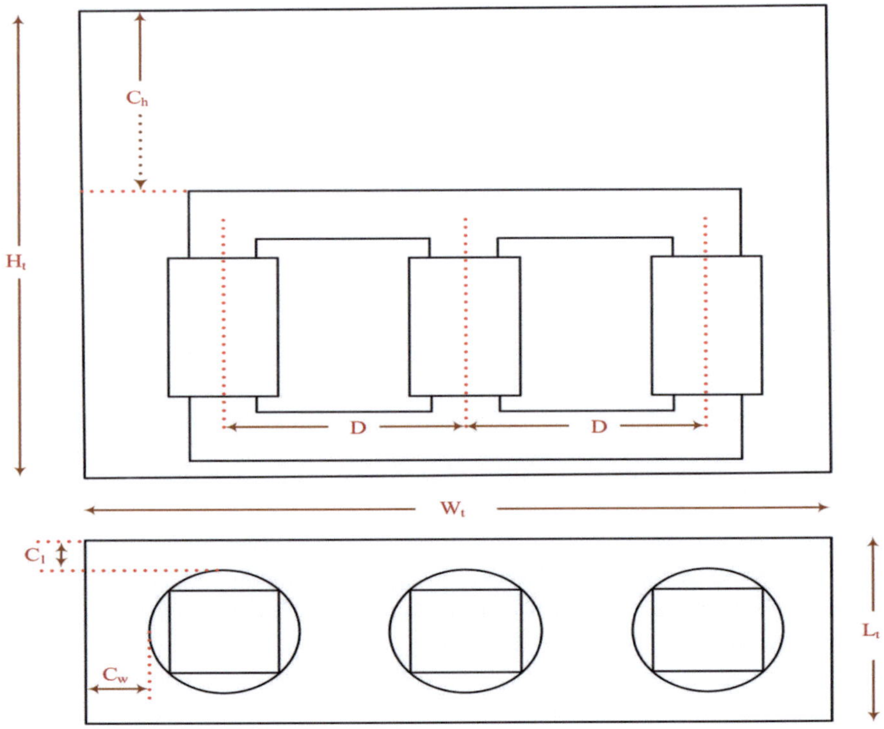


**Fig. 7.3** Transformer main dimensions with tank along with clearances

Tank's dimension = 1.25 × 1 × 0.5 m
We know that,

$$\text{Total Losses} = 12.5\, S_t\theta + 8.78\, A_t\theta$$

The top and bottom surfaces are not included, and the dissipating surface of the tank is accordingly;

$$S_t = 2H_t(L_t + W_t) = 2 \times 1.25(1 + 0.5) = 3.75\ \text{m}^2$$
$$5000 = (12.5)(3.75)(50) + (8.78)A_t(50)$$
$$5000 = 2343.75 + 439A_t$$
$$\Rightarrow A_t = 6.05\ \text{m}^2.$$

Let's assume that the diameter of the tube is 5 cm and the average height of the tube is 105 cm.

Each tube's dissipating area $a_t = \pi dl = \pi(0.05)(1.05) = 0.1649\ \text{m}^2$

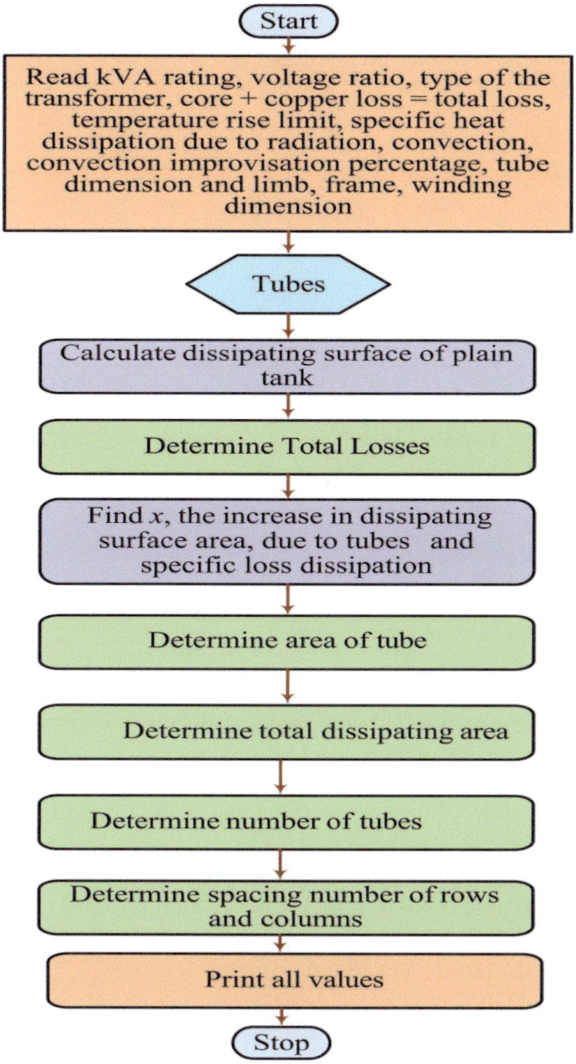


**Fig. 7.4** Flowchart for design of tank with tubes in transformer

$$\text{Number of tubes} = \frac{6.05}{0.1647} = 36.$$

When the tubes are placed across 7 cm distances from each other, there are 12 tubes on 100 cm side and 6 on the 50 cm side as depicted in Fig. 7.5.

$$\text{Total tubes } = (2 \times 12) + (2 \times 6) = 36$$

Specific heat of dissipation of a plain walled tank = $12.5/\text{W}/\text{m}^2/°\text{C}$

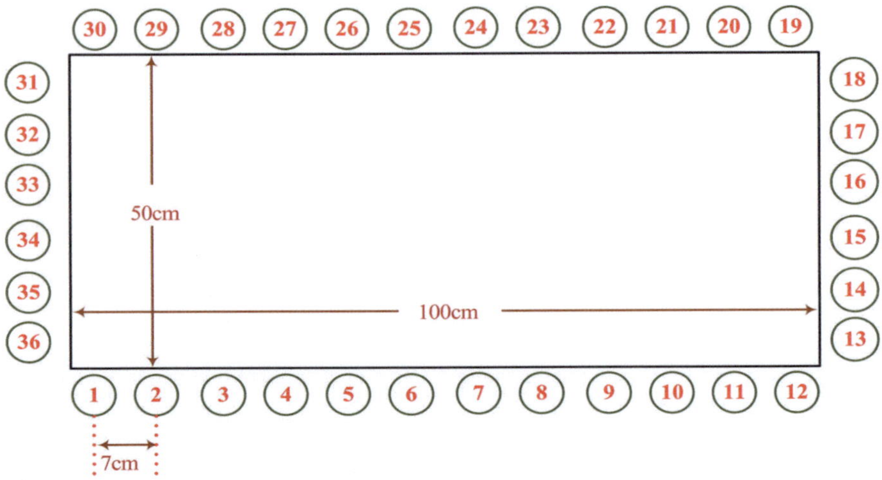


**Fig. 7.5** Cooling tubes arrangement for example 7.1

$$\text{Hence, the temperature rise of the plain walled tank} = \frac{5000}{12.5 \times 3.75}$$
$$= 106.66\ ^\circ\text{C}$$

An additional surface area of 36 tubes is provided to transfer heat in order to increase the temperature to 50 °C inside it.

***Example 7.2*** Design an efficient cooling system for a 500 kVA, 6600/440 V50 Hz 3ϕ transformer with a total load loss of 7 kW. The transformer has outside dimensions of 0.47 m in width, 0.96 m in length, and 1 meter in height. Cooling tubes OD should be 50 mm in order to maintain the increase of average temperature to no more than 35 °C. When it comes to the height, the length, and the width sides, you should leave 50 cm, 14 cm, and 13 cm, respectively.

**Solution**:

Given,
Transformer's rating, $Q$ and type = 500 kVA and $3\phi$
Voltage rating = 6600/440 V
Tank's dimension = 1 × 0.96 × 0.47 m
Clearances = 50 × 14 × 13 cm
Cooling tube's diameter = 50 mm
Acceptable temperature rise = 35 °C
The transformer's total size, comprising clearances on all sides,

$$H_t \times L_t \times W_t = (100 + 50) \times (96 + 14) \times (47 + 13)$$
$$= 150 \times 110 \times 60 \text{ cm.}$$

Total dissipating surface while neglecting the top and bottom surfaces

$$\begin{aligned}\text{Total dissipating surface while}&\\ \text{neglecting the top and bottom surfaces} &= 2[(H_t \times L_t) + (H_t \times W_t)]\\ &= 2[(1.5 \times 1.1) + (1.5 \times 0.6)]\\ &= 5.1\ \text{m}^2\\ \text{Total losses} &= 12.5S_t\theta + 8.8A_t\theta\\ 7000 &= (12.5)(5.1)(35) + (8.8)A_t(35).\end{aligned}$$

Tube's area,

$$A_t = 15.6\ \text{m}^2$$

Assuming the average height of each tube $= 0.9H_t = 0.9 \times 150 = 135$ cm
Each tube's dissipating surface area,

$$a_t = \pi Dl.$$

3.82 Design of Transformer

$$\begin{aligned} &= (\pi)(0.05)(1.35)\\ &= 0.212\ \text{m}^2\\ \text{Number of tubes required} &= \frac{A_t}{a_t}\\ &= \frac{15.6}{0.212}\\ &= 73.6 \simeq 74.\end{aligned}$$

That is, if the tubes are placed 7.5 cm apart, a number of tubes that can be placed along 110 cm and 60 cm sides are 110/7.5 and 60/7.5, respectively, which is equal to 15 and 8, respectively.

Total tubes used are 2(15+8) = 46 if 15 and 8 tubes are supplied. Extra 46 tubes can be used since 74 are required for heat dissipation; hence we have 92 tubes (46*2), which is way superior to 74.

Therefore, we can choose 13 tubes, which can be offered on 110 cm, and 6 tubes can be offered on 60 cm.

$$\text{Total number of tubes} = 2[13 + 6]+2[13 + 6] = 76$$

Therefore, as an alternative of 74 tubes, 76 can be used as shown in Fig. 7.6.

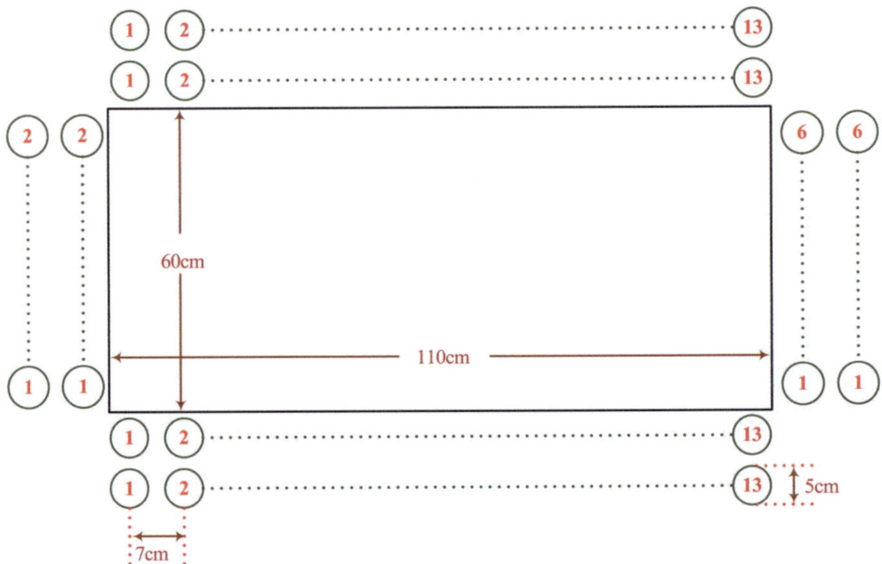


**Fig. 7.6** Cooling tubes arrangement for example 7.2

***Example 7.3*** Design a cooling system for a 1200 kVA natural oil-cooled operating transformer with a primary structure at 0.6 × 1.5 × 1.8 m. The transformer's total load loss is 12 kW. Our surface convection and radiation dissipation stand at 6 W/m²/°C and 6.5 W/m²/°C, respectively. The method with tubes boosts performance by 35%. The temperature can rise by forty degrees without problems. Use 5 cm long cooling tubes that measure 1 m in length and have a 5 cm diameter.

**Solution**:

Given

Rating of transformer = 1200 kVA

Main dimensions = 0.6 × 1.5 × 1.8 m

Full load loss = 12 kW

Allowable temperature rise = 40 °C

Improvement in convection = 35%

Loss dissipation due to radiation and convection = 6 $\text{W/m}^2/°\text{C}$ and 6.5 $\text{W/m}^2/°\text{C}$

$$\text{Total heat dissipating surface} = 2(H_t L_t + W_t H_t)$$
$$= 2 \times [(1.8 \times 0.6) + (1.5 \times 1.8)]$$
$$S_t = 7.56 \text{ m}^2$$

$$\text{Loss dissipation due to radiation and convection} = (6 + 6.5) S_t$$

$$= 12.5S_t$$

Heat dissipating area of tubes $= A_t$

Given that 40% improvement in dissipation due to convection is to be provided by tubes of area $A_t$.

Loss dissipated due to cooling tubes $= 6.5 \times \frac{135}{100} \times A_t = 8.77A_t$

Total losses dissipating area by tank and tubes $= 12.5S_t + 8.77A_t$

$$\begin{aligned}
\text{Total losses} &= 12.5S_t\theta + 8.77A_t\theta \\
12\ \text{kW} &= (12.5S_t + 8.77A_t)\theta \\
\theta &= \frac{12 \times 1000}{12.5S_t + 8.77A_t} \\
40 &= \frac{12 \times 1000}{(12.5)(7.56) + 8.77A_t} \\
A_t &= 23.43\ \text{m}^2
\end{aligned}$$

Total area of tubes,

$$A_t = 23.43\ \text{m}^2$$

3.84 Design of Transformer

$$\begin{aligned}
\text{Area of each tube} &= a_t = \pi \text{Dl} = (\pi)(5 \times 10^{-2})(1) = 0.157\ \text{m}^2 \\
\text{Total number of tubes} &= \frac{A_t}{a_t} = \frac{23.43}{0.157} = 149
\end{aligned}$$

Let 7.5 cm be the center-to-center distance between cooling tubes.

Number of tubes on 150 cm and 60 cm sides are $\frac{150}{7.5}$ and $\frac{60}{7.5}$, respectively.

So, 20 and 8 tubes can be placed. So, 20 and 8 tubes can be placed.

Number of tubes in each row $= 2(20 + 8) = 56$.

If two rows of tubes are placed, total number of tubes will be $2 \times 56 = 112$.

Since we have calculated only 150 tubes, the placement of tubes can be 20, 16, 20 it's widthwise and 8, 4, 8 along lengthwise as shown in Fig. 7.7.

Hence, the total will be $2(20 + 8) + 2(16 + 4) + 2(20 + 8) = 56 + 40 + 56 = 152$ tubes.

***Example 7.4*** Our design starts with a 3ϕ 10 MVA transformer that has 33KV and 6.6KV connections at 50 Hz. The transformer measures 250 cm high, 160 cm long, and 78.5 cm wide with distinct space requirements of 50 cm, 11.5 cm and 11.5 cm between each side for proper tank design. The whole transformer iron loss is 25 kW while the copper elements produce 100 kW losses.

Find out how much more the transformer's temperature will be or increase if cooling tubes are not installed. When the temperature is to be increased it is required

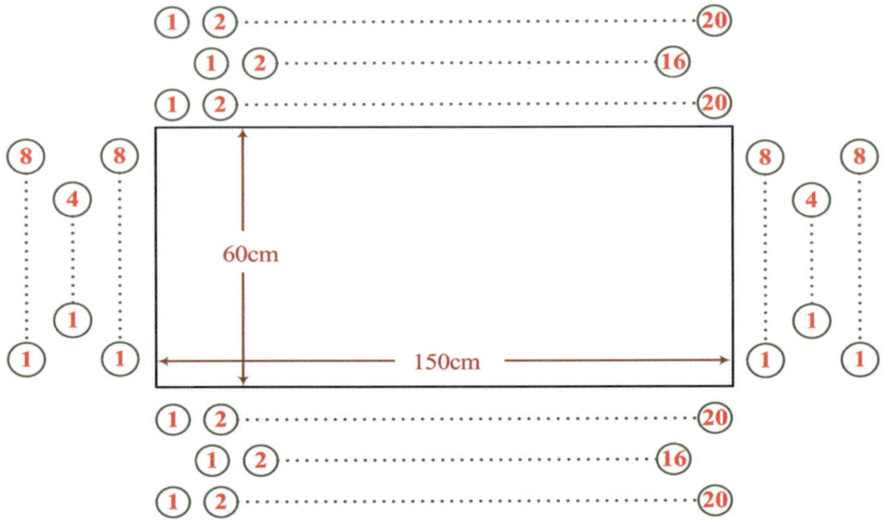


**Fig. 7.7** Cooling tubes arrangement for example 7.3

to keep the temperature increase to 50 °C to determine number of required tubes. Further, it is important to argue on the cooling system and on whether in a real system, cooling tubes should be used.

**Solution**:

Given

Transformer rating, $Q$ and type = 10 MVA Total iron loss = 25 kW and $3\phi$.

Voltage rating = $33,000/6600$ V

Allowable temperature rise = 50 °C

Total copper loss = 100 kW

Tank size = $1.60 \times 2.50 \times 0.785$ m

Clearances = $50 \times 11.5 \times 11.5$ cm

We know that

$$\begin{aligned}\text{Tank size} &= \text{Transformer size} + \text{clearances}\\ \text{Height} \times \text{width} \times \text{length of tank} &= (H + 50) \times (W + 11.5) \times (L + 90)\\ H_t \times W_t \times L_t &= 300 \times 90 \times 250 \text{ cm}\end{aligned}$$

Heat dissipating surface area of tank,

$$\begin{aligned}S_t &= 2H_t(L_t + W_t)\\ &= 2 \times 3(2.5 + 0.9)\\ &= 20.4 \text{ m}^2.\end{aligned}$$

Without cooling tubes,

$$\text{Total losses} = 12.5S_t\theta$$
$$125 \times 10^3 = (12.5)(20.4)\theta$$
$$\theta = \frac{125 \times 10^3}{(12.5)(20.4)} = 490\ ^\circ\text{C}$$

The temperature rise without cooling tubes is 490 °C. To limit this to 50 °C, the dissipating surface area has to be increased by providing cooling tubes. Let $A_t$ be the area of all cooling tubes.

$$125\ \text{kW} = 12.5S_t\theta + 8.78A_t\theta$$
$$125 \times 10^3 = (12.5)(20.4)(50) + (8.78)(A_t)(50)$$
$$A_t = 255.694\ \text{m}^2.$$

Let the average height of tube is $0.9H_t = 0.9 \times 300 = 270$ cm and the diameter of each tube is 5 cm.

Dissipating area of each tube $= \pi Dl = (\pi)(0.05)(2.7) = 0.424\ \text{m}^2$.

$$\text{Clearances} = 50 \times 14 \times 13\ \text{cm}.$$

## 7.9 Conclusion

The advent of temperature rise and its control form the very essentials of a power transformer's dependable performance, durability, and economy. High temperature adversely affects the insulation materials used for transformers, reduces the life expectancy of transformers small, and also increases the chances of failure. Solving this problem involves applying higher classes of design and employs such methods as airflow coolants and control of heat transfer. There are two main types of cooling for power transformers: natural and forced cooling, and each of them has its benefits depending on the size of the transformer, its operating conditions and application. Almost all distribution transformers are naturally cooled; oil-immersed natural air cooling (ONAN) is cheaper and commonly used on small transformers; forced cooling systems like oil-immersed forced air cooling (ONAF) and oil-immersed forced water cooling (OFWF) are required for big transformers. As for the primary cooling methods, there are additional smart technologies that advance dynamic parameters control depending on the real operative data of a forced cooling system.

Developing a suitable cooling system for power transformers requires a purely analytical approach based on thermal analysis, material selection, and system architecture. Using FEA and CFD, such as ANSYS and Fluent, computational analysis is used more often to analyze and optimize cooling systems using heat transfer and fluid flow. Sensors and automation are popular in large transformers and are now known as smart cooling solutions. It is worth stressing that the transformer tank plays an important part in the cooling process of a transformer. Cooled tanks, with the internal cooling tubes or radiators, increase the heat exchange area, therefore increasing the general cooling rate. The concept used when designing these tanks includes issues such as mechanical strength, thermal conductivity, and costs. Hyper specialization and high-tech thin applications on the thermo working components are also underdeveloped to enhance heat transfer coefficients and reduce the thermal insulation effects.

In conclusion, improvements in transformer cooling and the design of the tank have also improved the operating reliability of power transformers. The use of such modern technologies as smart cooling systems and new materials enabled the development of new possibilities of thermal management. These developments allow transformers to sufficiently meet the ultimate penetrating needs of the modern power systems with less adverse effects on the environment and operating expenses. Future work should be directed toward improvements of cooling technologies, eco-friendly materials, and automation technologies supporting the systems. Moreover, further usage of computational tools shall enable precise modeling and optimization of cooling systems as well as transformer tanks. These areas indicate the possible ways to enhance more significant performance and reliability of the power transformer industry.

## References


1. Patel D, Chothani N (2020) Digital protective schemes for power transformer, first. In: Power systems. Springer Singapore, Singapore. [Online]. Available: https://books.google.co.in/books?id=tjaSzQEACAAJ
2. Nilesh Chothani DP, Raichura M (2023) Advancement in power transformer infrastructure and digital protection, 1st edn. Studies in Infrastructure and Control, Springer, Singapore
3. Chothani N, Raichura M, Patel D (2023) Transformer infrastructure for power grid. In: Advancement in power transformer infrastructure and digital protection. Springer Nature Singapore, Singapore pp 1–26. https://doi.org/10.1007/978-981-99-3870-4_1
4. Chothani N, Patel D, Raichura M (2019) Transformer protection with sequence components and digital filters, 1st edn. LAP LAMBERT Academic Publishing, Latvia, European Union
5. Zhu B, Yang F (2021) Power transformer design practices, 1st edn. CRC press, Tailor and Francies group. https://doi.org/10.1201/9780367816865
6. Kulkarni SV, Khaparde SA (2004) Transformer engineering: design and practice. Taylor & Francis
7. Harlow JH (ed) (2012) Electric power transformer engineering, 3rd edn. CRC press, Tailor and Francies group. https://doi.org/10.1201/b12110
8. Vecchio RMD, Poulin B, Feghali PT, Shah DM, Ahuja R (2010) Transformer design principles: with applications to core-form power transformers, 2nd edn. Taylor & Francis. [Online]. Available: https://books.google.co.in/books?id=YDGHZ_acR-EC
9. Winders J (2002) Power transformers principles and applications, 1st edn. CRC press, Tailor and Francies group, New York. https://doi.org/10.1201/9780203910474

10. Flanagan WM (1993) Handbook of transformer design and applications. McGraw-Hill education. [Online]. Available: https://books.google.co.in/books?id=--lSAAAAMAAJ
11. López-Fernández XM, Ertan HB, Turowski J (eds) (2013) Transformers: analysis, design, and measurement, 1st edn. CRC press, Tailor and Francies group. https://doi.org/10.1201/b12275
12. McLyman CWT (2011) Transformer and inductor design handbook. CRC Press, Boca Raton. https://doi.org/10.1201/b10865
13. Dasgupta I (2012) Power transformers: quality assurance, 1st edn. New Age International
14. Behrooz Vahidi RAN (2023) Principles and modeling of the power transformers, 1st edn. Springer Singapore, Singapore. https://doi.org/10.1007/978-981-99-5796-5
15. Khan WA, Ellahi M, Masood B, Usman F, Abbass G (2018) Design of smart cooling system for power transformer. In: 2018 international conference on engineering and emerging technologies (ICEET), pp 1–7. https://doi.org/10.1109/ICEET1.2018.8338631
16. Goscinski P, Nadolny Z, Nawrowski R, Boczar T (2023) The tools and parameters to consider in the design of power transformer cooling systems. Energies 16(24). https://doi.org/10.3390/en16248000
17. Ngo M, Cao Y, Dong D, Burgos R, Nguyen K, Ismail A (2023) Forced air-cooling thermal design methodology for high-density, high-frequency, and high-power planar transformers in 1U applications. IEEE J Emerg Sel Top Power Electron 11(2):2015–2028. https://doi.org/10.1109/JESTPE.2022.3223849
18. Saket A, Pal S, Sharma AK (2018) Experimental study and performance analysis of transformer cooling system. Int J Res Trends Innov 3(9):91–98
19. Prajapati M, Pachori A (2015) Advance cooling technique of large power transformer and impacts of cooling on the age of transformer. Int J Electr Electron Res 3(2):651–661
20. Kim SE, Seop Ryu J (2023) A study on optimizing cooling design for a power transformer. In: 2023 29th international workshop on thermal investigations of ICs and systems (THERMINIC), pp 1–4. https://doi.org/10.1109/THERMINIC60375.2023.10325891
21. Sacco R, Carichino L, de Falco C, Verri M, Agostini F, Gradinger T (2014) A multiscale thermo-fluid computational model for a two-phase cooling system. Comput Methods Appl Mech Eng 282:239–268. https://doi.org/10.1016/J.CMA.2014.08.003
22. Patel D, Chothani N (2020) Real-time monitoring and adaptive protection of power transformer. In: Digital protective schemes for power transformer. Springer Singapore, Singapore, pp 173–190. https://doi.org/10.1007/978-981-15-6763-6_7
23. Chothani NG, Raichura MB, Patel DD, Mistry KD (2018) Real-time monitoring protection of power transformer to enhance smart grid reliability. In: 2018 IEEE electrical power and energy conference (EPEC). IEEE, Toronto, ON, Canada, pp 1–6. https://doi.org/10.1109/EPEC.2018.8598427
24. Raichura M, Chothani NG, Patel D (2019) Real-time monitoring and adaptive protection of power transformer to enhance smart grid reliability. Electr Control Commun Eng 15(2):104–112. https://doi.org/10.2478/ecce-2019-0014
25. Chothani N, Patel DD, Desai A (2023) Hysteresis current control-based boost rectifier analysis for single phase solid state transformer. In: 2023 international conference on recent advances in electrical, electronics and digital healthcare technologies (REEDCON), pp 385–390. https://doi.org/10.1109/REEDCON57544.2023.10151262
26. Patel D, Desai A, Chothani N (2023) Fixed frequency current control-based boost rectifier analysis for single phase solid state transformer. In: 2023 international conference for advancement in technology (ICONAT), pp 1–6. https://doi.org/10.1109/ICONAT57137.2023.10080301
27. Patel DD, Chothani N, Parmar BD (2023) Design and development of solid state transformer with different current control techniques. In: 2023 IEEE 11th region 10 humanitarian technology conference (R10-HTC), pp 414–419. https://doi.org/10.1109/R10-HTC57504.2023.10461788

# Chapter 8
# Computerized Design Methodology of Power Transformer

## List of Abbreviation

| | |
|---|---|
| LLC | Logically Link Control |
| SST | Solid State Transformer |
| MFT | Medium Frequency Transformer |
| THz | Tera Hz |
| AI | Artificial Intelligence |
| DLNN | Deep Learning Neural Network |
| MOG | Magnetic Oil Level Gauge |
| LVMB | Low Voltage Magnetic Balance |
| AT | Ampere Turn |
| NLL | No-Load Loss |
| LL | Load Loss |
| E.C | Electrolytic Copper |
| EHV | Extra High Voltage |
| UHV | Ultra High Voltage |
| CRGO | Cold-Rolled Grain Oriented |
| CCA | Core Coil Assembly |
| OLTC | ON Load Tap Changer |
| OCTC | OFF Circuit Tap Changer |
| CTC | Continuous Transposed Conductor |
| RIP | Resign Impregnated Paper |
| TDO | Transformer Design Optimization |
| HVMB | High-Voltage Magnetic Balance |
| LV | Low Voltage |
| HV | High Voltage |
| OTI | Oil Temperature Indicator |
| BDV | Breakdown Voltage |
| DVDF | Double Voltage Double Frequency |



N. Chothani et al., *Advancement in Infrastructure and Design Technology of Power Transformer*, Studies in Infrastructure and Control,
https://doi.org/10.1007/978-981-96-9420-4_8

Transformer design methodology in computers is an effective and comprehensive method of achieving optimal performance, cost-efficiency, and reliability with modern transformer designs. Essential calculations of the design framework for a step-by-step design are listed in this chapter, including the number of turns for HV and LV windings, core geometry, and overall transformer configuration. Core step design and winding gradients are designed carefully so that magnetic balance and minimal stray losses are achieved. The leakage flux is ensured to be within the acceptable limit, and magnetic efficiency is raised at the expense of ampere-turns (AT) balance. Insulation and mechanical stability standards are met through key performance parameter evaluation, which includes capacitance, impulse withstand strength, inrush current and dynamic short circuit resistance. An assessment of a transformer's ability to handle mechanical and thermal stresses during operation is made by calculating radial bursting force and thermal capacity. Performance and economic optimization also give an indication of efficiency, voltage regulation, as well as the total cost of the active part. Furthermore, tank design, pressure handling, vacuum withstand, oil requirements, as well as cooling system design, radiator selection, are calculated. All of these elements are put together into a MATLAB-based program that integrates and optimizes the transformer design process, resulting in rapid prototyping and accurate analyses for the standard and custom application.

## 8.1 Introduction

The history of transformer design began in the 1830s with the research conducted by Joseph and Faraday on electromagnets, which improved the inductor's properties when two windings were mutually coupled. Nicholas took things a step further and created the induction coil in 1836. Y. Pavel then employed these coils for the lighting in 1876. After the induction coil was developed and used in the electrical system concept of the transformer design in 1885, the Gang Company in Budapest developed the condensate light based on the AC supply between 1878 and 1883. William Stanley created the first commercial transformers in 1886 under certain working conditions. Because of its sophisticated internal architecture and variable characteristics, transformer infrastructure is particularly intriguing to study in depth [1–3].

A low inductance transformer was tested on an LLC-based converter to reduce leakage inductance, considering losses and varying resonant frequencies [4]. To effectively calibrate the conventional transformer with digital sampling multimeters for transformers rated up to 10kA current [5]. A transformer with a distinct frequency was devised to achieve size reduction and enhance its multifunctionality, based on numerous concepts, referred to as SST [6–8]. Currently, high insulation is applicable to 100 kW transformers on MFT, which exhibit notable efficiency and enhanced thermal performance [9]. It is important to note that the transformer heat sink bobbin material is utilized in three printed layers [9]. A unique transformer is engineered for electric car charging, utilizing a three-winding transformer in conjunction with a three-series resonant converter circuit. This circuit offers electrical isolation through power electronic circuits utilizing leakage inductance [10]. Although these circuits have been validated on hardware prototypes, significant testing for the various leakage inductances in the converter is still necessary, as noise substantially affects its characteristics. It is necessary to eliminate irrational sounds or to redistribute those sounds in various manners [11]. A high-frequency transformer significantly decreases the size of the transformer. Nonetheless, the exceptionally high frequency inside the THz range demonstrates that the polarization ratio reaches 91%, while simultaneously enhancing cross-polarization to over 50% [12]. This methodology enhances tools for communication, image processing methods, and the design of medical equipment.

Currently, several algorithms are employed to enhance transformer design. The heuristic algorithm serves as an example for the construction of distribution transformers [13]. This technology is evaluated against other ways based on enhanced efficiency and decreased transformer weight percentage. The converter transformers employ optimization strategies utilizing artificial intelligence and deep learning neural networks [14]. The suggested system was evaluated using FEM, ANSYS, and OPEL-RT software. The majority of wind farms are situated near coastal regions or in elevated terrains. This indicates that the offshore region is factored into the wind farm's assessment, as temperature significantly influences accuracy. It is expected to favor a high-temperature superconducting transformer over a normal transformer [15]. To minimize the leakage impedance of the dry-type transformer, it is recommended to enhance the gap positions utilizing optimization approaches [16]. The saturation of the transformer core results in the generation of inrush current. A unique method for limiting inrush current through core saturation is proposed. A claimed approach is improved by diminishing the transformer's axial and radial forces by 90% and 80%, respectively [17]. The DC fading current is often produced at various fault initiation angles. The addition of DC bias current enhances transformer performance and reduces failure analysis [18]. It also substantially impacts transformer design concerning inrush characteristics and improves the design parameters of protective schemes from a protection perspective [19–22].

This chapter addresses design parameters, computer-aided design, sequential design steps, performance parameter calculations, core step calculations, winding gradients, loss calculations, AT calculations, capacitance, impulse, inrush, dynamic short circuit, radial bursting force, thermal capacity, efficiency, regulations, tank calculations, pressure vacuum, and oil calculations.

It is important to design the particular equipment based on which place where it is going to be utilized. As per the commissioning space importance, designing parameters are set by the designing engineers. Improving compatibility, with a reduction in complexity and cost, is the main motto of the designers. Optimization of the different parameters is affected by the required specifications of the clients. It helps to improve the design of electrical as well as magnetic circuits in a transformer. There are many efforts placed into designing the power transformer based on trial and error considerations [23] based on condition monitoring of the transformers. Many efforts are taken based on numerical and analytical processes to develop proper procedures regarding to design of the transformers. Due to the advanced technology and computational techniques available in the market, the design process has become smoother optimizing the system parameters as per the consumer's requirement. So, it also reduces the complexity of the logic in terms of nonlinearity of the core and geometrical complexity as per the static device characteristic. Different thermal, chemical, electromagnetic, mechanical, and other parameters/characteristics are very sharply included in those computerized numerical techniques. After appearing as an artificial intelligence (AI) in the picture of engineering, different algorithm applications with optimum design parameters are obtained with asset management of the transformer unit. It is noted that the design of the transformer is very unusual, as per the application of the transformer, like welding transformer, power transformer, current transformer, potential transformer, etc. The parameter range also varies drastically with the use of those transformers. Cooling techniques, specific material characteristics accounting for its dynamic behaviors, and many more [23, 24] things are considered for the design of the transformers. Most of the transformer research emphasized more on technical characteristics like NLL & LL and leakage, field, inrush current, dynamic behavior, noise, insulation, cooling, etc.

Cost optimization is the main objective of the designers, including all required features and equipment performance as per the demand of the client. For the equipment performance, it is compulsory to evaluate the way and causes of the losses. All those parameters directly affect the design system, constraints, and resolution procedures. So, it becomes more and more complicated to deal with the system design. It is also desired to develop a system with cost-effectiveness including flexibility. Here, in this chapter research is associated with dealing the parameter optimization including different techniques of the mathematical process. It is also very imperative data in any optimization practices concerning its efficacy and challenging investigation capability. It is the final goal to resolve the problem formulation with a complete informative form. As far as concerns, the well-cultured transformer designing, it must satisfy different constraints like thermal, chemical, mechanical, electrical, and other technical parameters. Here, the transformer design includes optimization procedures dealing with the operating and manufacturing costs in terms of utility and system. Adopted techniques are considered here as a multiple design technique. These techniques involve different and enormous substitute parameters to design to cultivate massive alternative designs. In the end, it is the main requirement of the design to fulfill all constraints of the said problem, including the lowest establishing cost [25].

## 8.2 Computer-Based Transformer Design Methodology

The transformer design can be very efficiently carried out by a computer. The major design factors, like iron loss, load loss, % impedance, and finally the cost of the active materials, are calculated. The calculations are based on the same assumptions as in the classical theory, i.e., the procedure followed by a designer, the core diameter, core area, core loss, LV winding design, HV winding design, and % impedance calculation, and so on. The laborious and time-consuming design calculations are carried out by the computer programs in a few seconds. The computer results are also, without exception, accurate. However, the computer lacks the judgment of a designer, and hence every step in the design process has to be investigated and written down exactly because the computer cannot take account of all the various factors influencing it at the same time.

The various computer programs help the designer to select types of windings, inter disk blocks, conductor coverings, inter winding ducts, net core area, core construction, short circuit forces that are likely to come, thermal design and radiator requirement, estimate eddy, and stray losses, estimate impulse voltage that is likely to come on the disks and whether inter disk block and conductor covering can withstand reliably the impulse voltage.

For transformer design programs, the following key parameters are operated on various ranges, which enables hundreds of design executions in a short time aimed at achieving an optimum design that provides the best efficiency and low cost.

It is the main requirement of the transformer design that, designer must consider the various parameters and different technical constraints while aiming for optimum design. A few of them are listed in Table 8.1.

- Constraint of No-Load and Load Loss.
- Constraint of Stray Loss.
- Constraint of Voltage Stress and Inrush Current.
- Constraint of Short Circuit Forces.
- Constraint of Insulation Coordination.
- Constraint of Heat Dissipation.

The design procedure asks for consideration of several assumptions for low-cost and highly efficient design. 70% of the total transformer cost is covered by two major materials, namely copper and lamination. All above said variables are associated with each other, like conductor geometry, which has a great influence on load loss and %Z, while flux density influences the core geometry and finally, on no-load loss. All procedures for designing the transformer are followed as per Fig. 8.1.

**Table 8.1** Variables, monitoring, and output parameters of transformer design

| Variable parameter | Monitoring parameter | Output parameter |
|---|---|---|
| K-factor | Core and winding geometry | Key-raw material cost |
| Conductor geometry | Current density | No-load loss |
| Conductor arrangement | Flux density | Load loss |
| CRGO material grade | Temperature gradient | % impedance |
| Clearances | | Temperature rise |

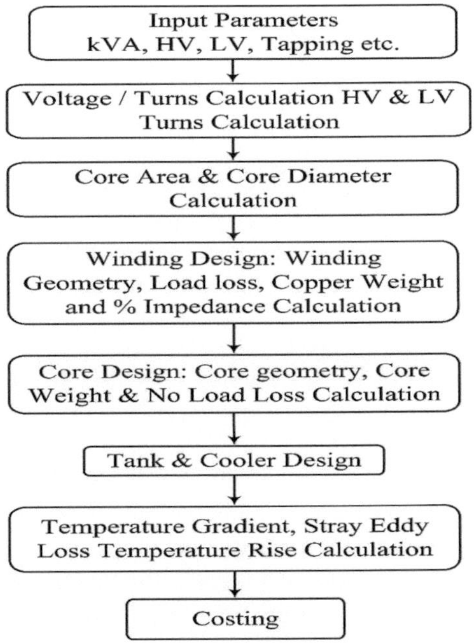


**Fig. 8.1** Transformer design process flowchart

The flux density ($B_m$), current density, and $K$-factor are the parameters that influence greatly to the customer's requirements, viz., no load, loss, load loss, %Z, and cost too. The interrelation between these parameters is shown in Fig. 8.2.

Apart from the mathematical equations, the transformer design process asks for several standard data and assumption factors, which are also described herewith.

## 8.3 Steps in the Sequential Design Process of a Transformer

The design sequence with respective formulas for a three-phase core-type transformer is as follows.

### *8.3.1 No. of Turns Calculation*

Here, the kVA rating is used to calculate per-turn emf. So, per turn, the emf is intended using the given equation,

$$E_t = K\sqrt{Q}. \tag{8.1}$$

Here, $Q$ = Transformer kVA

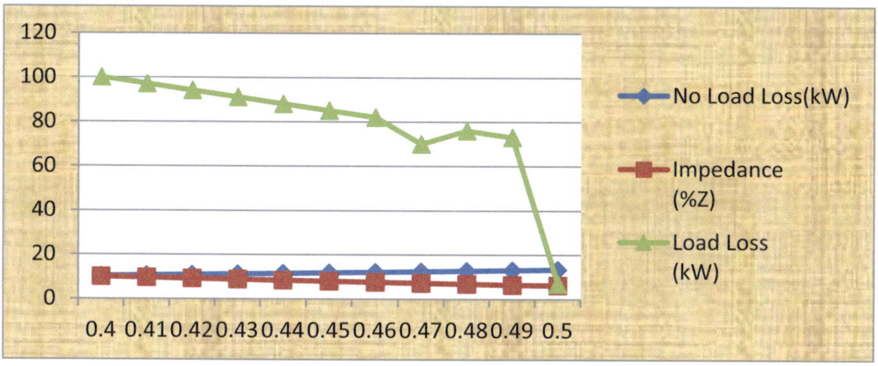


(A)

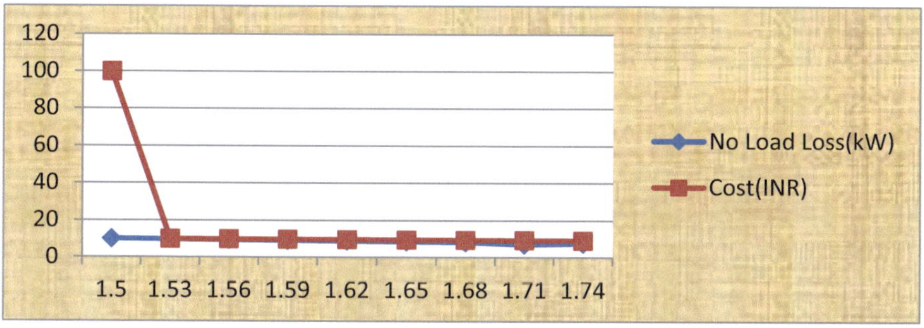


(B)

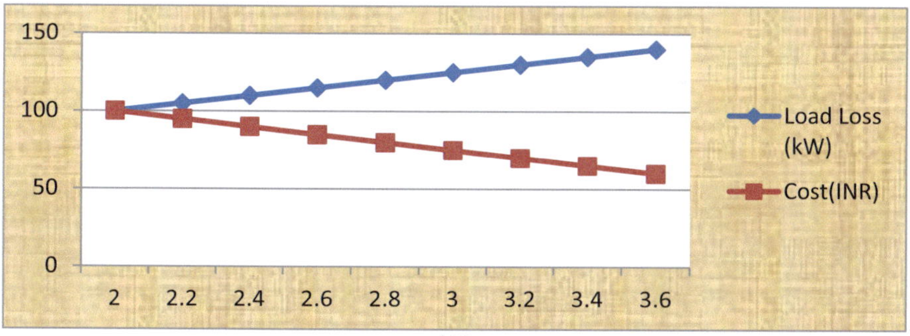


(C)

**Fig. 8.2** Parametric impact on the cost and losses of the transformer **a** NLL, load loss, and %Z comparison, **b** NLL versus cost of transformer, **c** load loss versus cost of transformer

Primary & secondary turn numbers are considered as trails equations,

$$T_{HV} = V_{HV}/E_t \tag{8.2}$$

$$T_{LV} = V_{LV}/\left(\sqrt{3} \times E_t\right) \tag{8.3}$$

### 8.3.2 Core Design

Here, core diameter, core area, and necessary core steps are taken into consideration when designing the transformer's core. The density of the flux within the core, its per-turn emf, and its frequency all influence its design.

The area of the core is calculated as follows,

$\mathrm{Ai} = \frac{E_t \times 10^4}{4.44 \times B_m \times f}$ Core diameter is being calculated as follows,

$$\mathrm{d} = \sqrt{\left(\frac{\mathrm{Ai}}{0.69}\right)} \times 10. \tag{8.5}$$

The diameter of the core is obtained from Eq. (8.5), taking a round-off to the nearest value, the number of steps is calculated.

If, $N_s = .\ (\mathrm{d} \times 0.04) > 17$ then $N_s = 17$ otherwise take $N_s = \mathrm{d} \times 0.04$.

### 8.3.3 Design of HV and LV Winding

As far as a concern in transformer design, there are two main windings present, referred to as HV and LV windings. It is most commonly a requirement to design a transformer winding. It should include different parameters like the rated current of the winding, winding cross-sectional area, no's of disks involved in winding, required tapping on each disk, the number of turns required per disk, and axial height during un-shrunk and under shrunk conditions. Both the inner and outer diameters of the winding are calculated as per some international or national standards.

$$A_{\mathrm{lvcu}} = I_{\mathrm{LV}}/\delta_{\mathrm{LV}} \tag{8.6}$$

$$\mathrm{LVh} = \mathrm{To}_{\mathrm{disk}} \times H_c + ((\mathrm{Todisk} - 1) \times \mathrm{Indisk}) + \mathrm{Intap} \tag{8.7}$$

$$D_{\mathrm{lvr}} = C_t \times N_{\mathrm{pc}} \times T_{\mathrm{pd}} \times 1.015 \tag{8.8}$$

$$\mathrm{LV}_{\mathrm{id}} = \mathrm{d} + 2 \times C_{\mathrm{lvc}} \tag{8.9}$$

$$\mathrm{LV}_{\mathrm{idm}} = \mathrm{LV}_{\mathrm{id}} + D_{\mathrm{lvr}} \tag{8.10}$$

$$\mathrm{LV}_{\mathrm{od}} = \mathrm{LV}_{\mathrm{idm}} + D_{\mathrm{lvr}} \tag{8.11}$$

Equally calculation for the HV winding.

## 8.4 Calculation of Performance Parameter

After a winding design, some performance parameters like percentage impedance, core loss, the weight of the core, copper, load losses, and percentage efficiency are calculated.

$$\%Z = \sqrt{(\%R_2 + \%X_2)} \tag{8.12}$$

$$W_C \frac{(7.65 \times (Hw \div 10) + (4 \times clc \div 10) + (2.1 \times d \div 10))}{1000} \times Ai \tag{8.13}$$

$$\text{Closs} = W_c \times f_f \times M_F \times \text{Watts/kg} \tag{8.14}$$

$$W_{lvcu} = \frac{\text{LV}_{id} \times \pi \times A_{lvcu} \times T_{lv} \times 8.89 \times 3}{1000000} \tag{8.15}$$

$$W_{hvcu} = \frac{\text{HV}_{id} \times \pi \times A_{hvcu} \times \text{T}_{hv} \times 8.89 \times 3}{1000000} \tag{8.16}$$

$$\text{LV}_{culoss} = W_{lvcu} \times \delta^2\text{LV} \times 2.4 \tag{8.17}$$

$$\text{HV}_{culoss} = W_{hvcu} \times \delta^2\text{HV} \times 2.4 \tag{8.18}$$

$$\text{Culoss} = \text{LV}_{culoss} + \text{HV}_{culoss} \tag{8.19}$$

$$\text{Toloss} = \text{C}_{loss} + \text{Cu}_{loss} + (0.2 \times \text{Cu}_{loss}) \tag{8.20}$$

$$\%\eta = \frac{\text{Output power}}{\text{Output power} + \text{Toloss}} \tag{8.21}$$

### *8.4.1 The Total Cost of the Active Part*

$$C_{cost} = W_c \times (\text{material cost/kg}) \tag{8.22}$$

$$\text{LVc}_{cost} = W_{lvcu} \times (\text{material cost/kg}) \tag{8.23}$$

$$\text{HVc}_{cost} = W_{hvcu} \times (\text{material cost/kg}) \tag{8.24}$$

## 8.5 Core Step Calculation

Different methods for calculating the core:

Step Count: = (core diameter *0.04) > 17; if not, then 17 (core diameter *0.04).

Core length is equal to (3 * leg center/10 + 4 * window height/10 + 2.1 * core diameter/10).

((core diameter)$^2$ – ((core diameter$^2$ – (core diameter * 0.2)$^2$)$^{0.5}$ – (core diameter * 0.027))$^2$)$^{0.5}$ is the constant (G). {Take note: Round the value (*by 5 & / by 5)}.

Step width minimum: 50 if G is less than 50; else, G

First-step maximum width = (core diameter * 0.98). {Take note: Round the value (*by 5 & / by 5)}.

Step 2: Width = Maximum (10) {Note: (10–15–25 progressively)}.

Step No. 1 thickness = ((step 1 width)$^2$)$^{0.5}$ – (core diameter)$^2$

Step No. 2 thickness = ((Second width)$^2$)$^{0.5}$ – (Core diameter)$^2$ (thickness of 1st step).

Thickness = ((Core diameter)$^2$) in Step No. 3 (step 3 width).

(2) 0.5 – (first step thickness plus second step thickness).

Area of ($N$ step) equals $N$ step thickness * $N$ step width * 0.0095.

Area of second step = width of second step – thickness of step * 0.0095.

{Note: Multiplying factor: 5 if the core diameter is less than 300, 10 otherwise} for the initial stack width.

The sum of all step areas equals the total area. \Note: (0.1*core diameter) if the core diameter is more than 390, otherwise 0}.

## 8.6 Winding Gradient and Stray Loss Calculation

### 8.6.1 LV Winding Gradient

Winding surface: ((MLT)*(LV Wdg Radial + Conductor area) * 2 * Nos of Disk * 0.67/10$^6$).

Eddy loss:

$$\begin{aligned}&\left(10 * \left((\text{Nos. of Layer } + \text{covered conductor width})^2 - 0.2\right)\right.\\&\quad * (\text{cond. Bare Width}/10)^4 * ((\text{Nos}\\&\left.\text{of Disk} + \text{Overall cond. Width})/\text{Wdg Axial})^2 * (\text{freq}/50)^2 + 1\right)\end{aligned} \tag{8.26}$$

$$\text{W/m}^2 = \left(I^2R + \text{Eddy} + \text{Connection loss }\right)/3/\text{Wdg surface}$$

$$\text{Gradient} = \left((\text{w/m}^2)/100 + (\text{w/m}^2)/200 * \text{ Cond. Covering }/2\right) \tag{8.27}$$

$$\begin{aligned}\text{Top oil rise} = \text{If }&(\text{Guaranteed Wdg rise} - \text{gradient})/0.8\\ &> (\text{top oil rise guaranteed} - 2),\\ &(\text{top oil rise guaranteed} - 1),\\ &(\text{Guaranteed Wdg rise} - \text{gradient})/0.8)\end{aligned} \tag{8.28}$$

LV Wdg Rise: (Top oil rise* 0.8) + gradient.
HV Wdg Rise: (Top oil rise* 0.8) + gradient.

### 8.6.2 HV Winding Gradient

Winding surface: ((MLT)* Wdg Radial *2*(Nos of Disk Normal + Tap Disk/2)*0.67/10^6)

$$\begin{aligned}\text{Eddy loss}: \Big(&10 * \big((\text{Nos. of turns per disk} * \text{Nos of parallel cont.})^2 - 0.2\big)\\ &* (\text{cond. Bare Width}/10)^{4*}((\text{Nos of Disknormal}\\ &* \text{Axial conduct dimension} + \text{Nos of Disk tap}\\ &*\text{Axial conduct dimension})/\text{Wdg Axial})^2\\ &* (\text{freq}/50)^2 + 1\Big)\end{aligned} \tag{8.29}$$

$$\text{W/m}^2 = \big(\text{I}^2\text{R} + \text{Eddy} + \text{Connection loss}\big)/3/\text{Wdg surface}$$

$$\text{Gradient} = \big((\text{w/m}^2)/200 + (\text{w/m}^2)/100 * \text{Cond. Covering}/2 * 1.1\big) \tag{8.30}$$

### 8.6.3 Stray Loss (LV)

$$\big(1200 * (E_t * \%X/222)^{1.7*}(2000/\text{Wdg Axial Height})^{0.5} * (\text{KVA}/1000/100)^{0.25} \tag{8.31}$$

### 8.6.4 Stray Loss (HV)

$$\big(\text{LV } I^2R + \text{LV Eddy} + \text{LV Connection Loss} + \text{LV Stray Loss} + \text{HV } I^2R + \text{HV Eddy}\big) \tag{8.32}$$

### *8.6.5 Current and Flux Density Calculation*

$$\text{Flux Density} = \text{LV Volt}/\left(\sqrt{3} * \text{LV Turns} * \text{Core Area} * .0222\right)\text{T} \tag{8.33}$$

## 8.7 Ampere Turns (AT) Balance Calculation

| HV | | | LV | | |
|---|---|---|---|---|---|
| Winding portion Ht | $D_1$ | *** Wdg height $Ht_1$ | $ht_1$ | Winding Portion Ht | $d_1$ |
| Phase current | $A_1$ | | | Phase current | $a_1$ |
| Nos of disks | $B_1$ | | | Nos of disk | $b_1$ |
| Nos of turns | $C_1$ | | | Nos of turns | $c_1$ |
| $AT_1$ | $A_1 * C_1$ | | | at1 | $a_1 * c_1$ |
| $AT_1$/MM | $(A_1*C_1)/D_1$ | | | at1/MM | $(a_1*c_1)/d_1$ |
| Winding portion Ht | $D_2$ | $Ht_2$ | $ht_2$ | Winding ortion Ht | $d_2$ |
| Phase current | $A_2$ | | | Phase current | $a_2$ |
| Nos of disk | $B_2$ | | | Nos of disk | $b_2$ |
| Nos of turns | $C_2$ | | | Nos of turns | $c_2$ |
| $AT_2$ | $A_2 * C_2$ | | | $at_2$ | $a_2 * c_2$ |
| $AT_2$/MM | $(A_2*C_2)/D_2$ | | | $at_2$/MM | $(a_2*c_2)/d_2$ |
| Winding portion Ht | $D_3$ | $Ht_3$ | $ht_3$ | Winding portion Ht | $d_3$ |
| Phase current | $A_3$ | | | Phase current | $a_3$ |
| Nos of disk | $B_3$ | | | Nos of disk | $b_3$ |
| Nos of turns | $C_3$ | | | Nos of turns | $c_3$ |
| AT 3 | $A_3 * C_3$ | | | at3 | $a_3 * c_3$ |
| AT3/MM | $(A_3*C_3)/D_3$ | | | $at_3$/MM | $(a_3*c_3)/d_3$ |
| Total(HT(hv)) | $(Ht_1 + Ht_2 + Ht_3)$ | | Total (HT(lv)) | | $(ht_1 + ht_2 + ht_3)$ |

| Total no. of disks | Disk HV | Disk LV |
|---|---|---|
| Total no. of turns | $(C_1 + C_2 + C_3)$ | $(c_1 + c_2 + c_3)$ |
| Total AT: | $At_1 + At_2 + AT_3$ | $at_1 + at_2 + at_3$ |
| Axial Ht.: | HT(hv) | HT(lv) |
| AT/MM: | AT(HV)/HT(hv) | at(lv)/Ht(lv) |

***Winding portion Ht = (Nos of disk x Conductor ht x Nos axially conduct) + (0.8 × Nos of disk × Nos of axially cond × 0.85) + ((Nos of disk-1) × spacer thickness × 0.95)

## 8.8 Capacitance Calculation

### 8.8.1 Parallel Capacitance

$$\begin{aligned} C(1\&2) &= \left(D^{*}H\right)/36/((X/2.7) + (Y/4.2)) \\ C &= C_1(\text{HV}) + C_2(\text{LV})\text{pF}, \end{aligned} \tag{8.34}$$

where

D Mean diameter of winding considered.
X Thickness of oil insulation bet wdg (mm) (duct).
Y Thickness of solid insulation (PB) between wdg (mm) (cylinder).
H Axial wdg. Length (mm).

### 8.8.2 Series Capacitance

$$\begin{aligned} C_{\text{si}} &= \left(3.2^{*}D^{*}(W + (2 * t))^{*}(n - \text{d} - 1)\right)/\left(36^{*}t^{*}M\right) \ \text{pF} \\ C_{\text{sn}} &= (D/36/M) * (\text{A} + \text{B} + \text{C})\text{PF} \\ A &= 3.2 * (h + (2 * t)) * (m - \text{d} - 1)/m^2/t \\ B &= \left(h + t + g'\right)^{*}\text{d}/m^2/((g_{\text{a}}/2.7) + (t/3.2)) \\ C &= 4/3 * (R + g_r)/((t/3.2) + (g_r/2.8)), \end{aligned} \tag{8.35}$$

where

*D* Mean diameter of winding in mm.
*W* Bare conductor axial width.
*t* Conductor covering.
*n* Number of layers.
*D* Number of ducts between layers.
*M* Number of Turns.
*M* Number of conductor per layer (axially parallel).

$g_a$ Axial Duct width.
$g_r$ Radial Spacer thickness.
$g'$ Smaller of $g_a/g_r$.
$h$ Bare conductor width.
$r$ Radial depth.

## 8.9 Impulse Calculation

Continuous Disk/Helical Winding.

$L$ Physical length of winding.
$x$ Cu to Cu dist. bet. two line disks/turns.
$n$ No. of disks.
$d$ Radial dist. HV to LV.
$s$ Rad. depth of winding

$$\alpha = \sqrt{\frac{L * X * n}{d * s}} \tag{8.36}$$

## 8.10 Inrush Current Calculation

$$\text{Io max} = H_w \times A_i \times (2 \times B_m + B_r - 2.2)/N \times A_c \times \mu 0 \tag{8.37}$$

where

$H_w$ wdg ht.
$A_i$ Core area.
$B_m$ Flux density.
$B_r$ Remnant flux density(approx. 60% of $B_m$).
$N$ No. of wdg. turns.
$A_c$ Mean area enclosed by winding.
$\mu_0$ Air permeability ($4\,\Pi \times 10^{-7}$).

## 8.11 Dynamic Short Circuit Withstand Capacity Calculation

Dynamic Ability Calculations for Short Circuit.

### 8.11.1 Computing the Asymmetrical Short Circuit Current's First Peak Value

$$\text{Isc} = k^{*}\text{Sqrt}(2)^{*}I_{\text{ph}}/I_{\text{z}} \tag{8.38}$$

where $k$*sqrt(2) is a function of $x/r$

$$k^{*}\text{sqrt}(2) = 1.8 * 1.414 = 2.55$$

$I_{\text{ph}}$ Rated phase current in Amps.
$I_{\text{z}}$ Per unit impedance.

### 8.11.2 Asymmetrical Short Circuit Ampere Turns Calculation

$$\text{Amp turns} = N * I_{\text{sc}} \tag{8.39}$$

where $N$ = No. of turns

$I_{\text{s}}$ Asymmetrical short circuit current.

### 8.11.3 Hoop Stress

$$\text{HP(mean)} = K * I_{\text{ph}} * I_{\text{ph}} * R_{\text{dc}}/(H_{\text{t}} * I_{\text{z}} * I_{\text{z}})\text{Kg/Cm}^2 \tag{8.40}$$

where

$K$ 0.03 for copper.
$R_{\text{dc}}$ DC resistance of the phase at 75 °C in Ohms.
$H_{\text{t}}$ Height of the winding in cm.
$P_{\text{avg}}$ HP (HV) + HP (LV)/2.

The permissible limit is 1250 kg/cm$^2$

## 8.12 Radial Bursting Force

$$F_{\text{r}} = 2 * \Pi * H_{\text{p}}\text{mean} * I_{\text{ph}} * N/\text{Density Kgs} \tag{8.41}$$

where density = Current density in Amp/cm$^2$
Hence,
No. of supports to be provided in the winding

$$N_{\text{s}} = (D_{\text{mi}}/B_{\text{i}}) * \text{sqrt}(12 * \text{HP avg/E})$$

where

$E$ Elasticity constant = 1.13 * 106 kg/cm$^2$
$D_{mi}$ Mean diameter on the side of the winding in cm.
$B_i$ Thickness of inside winding conductor in cm.

### 8.12.1 Internal Axial Compression

$$F_c = -34 * S_n/(I_z * \mathrm{Ht})\mathrm{Kg} \quad (8.42)$$

where

$S_n$ Rated kVA.
(−) Indicates that the force is acting toward the center.

### 8.12.2 Axial Imbalance Force Calculation Owing to Tapping in the Winding

$$F_a = (a/2) * (N * I_{sc})^2 * 10 - 7 \quad (8.43)$$

where a = per unit turns out of circuit in the winding = 0.05

The force will be half of the amount computed above because of the supplied compensatory gap. Hence, $F_a = F_a/2$ kg.

### 8.12.3 Maximum Compressive Pressure on Spacers

$$P_i = (F_a + (2/3 * F_c))/A_i \quad \mathrm{Kg/cm^2} \quad (8.44)$$

$$P_o = (F_a + (1/3 * F_c))/A_o \quad \mathrm{Kg/cm^2} \quad (8.45)$$

where

$P_i$ Compressive pressure in the inner radial spacers in Kg/cm$^2$
$P_o$ Compressive pressure in the outer radial spacers in Kg/cm$^2$
$A_i$ Total supported area of the inner radial spacer in cm$^2$
$A_o$ Total supported area of the outer radial spacer in cm$^2$.

The permissible limit is 500 kg/cm$^2$.

### 8.12.4 Tensile Stress in Tie Rods/Flitch Plates

$$P_t = (F_a - (F_c/3))/\text{Total Area Kg/cm}^2 \quad (8.46)$$

where Total area = No. of tie rods or flitch plates * size
Which is well below the permissible limit of 2500 kg/cm$^2$.

### 8.12.5 Bending Stress on Clamping Structure

$$\text{Bs max} = (6 * \Pi * F_a * \text{D})/\left(8 * b * t^2 * n^2\right) \quad (8.47)$$

where

D Diameter of the ring in cm.
b Width of the ring in cm.
t Thickness of the ring in cm.
n No. of jacking pads.

## 8.13 Thermal Ability Calculation

$$\text{Symmetrical short circuit current}(I) = \text{UKA}(Z_t + Z_s) * 1.73.$$

where

U Rated voltage of the winding.
$Z_t$ (Short circuit impedance of the transformer) = $U_z * U_n^2 100 * S_n$.
$U_z$ % Impedance.
$U_n$ Rated voltage.
$S_n$ Rating in MVA

$$\text{Final temperature}(\theta_1) = \left(\theta^0 + a^* j^2 * t + *10^{-3}\right) \quad (8.48)$$

where

$$\text{Initial temp}(\theta_0) = 10^5,$$

$$\text{Current density under S.Ckt(J)} = \text{Sym S.Ckt current/copper area}$$

$$\text{Short circuit duration(t)} = 2\,\text{Sec}$$

$$a = 8.2$$

$$\text{Permissible limits} = 250\ ^{\circ}\text{C}.$$

## 8.14 Efficiency and Regulation Calculation

$$\begin{aligned}\text{Efficiency } \% = {} & \text{KVA} \times \text{P.F} \times \%\ \text{Load} \times 100\{(\text{KVA} \times \text{P.F} \times \%\ \text{Load}) \\ & + (\text{N.L.L. Loss(KW)}) + ((\%\ \text{Load})\, 2 \times \text{Load Loss (KW)})\}\end{aligned} \tag{8.49}$$

$$\begin{aligned}\text{Regulation } \% = {} & (\%\ \text{Resistance} \times \cos \emptyset) + (\%\ \text{Reactance} \times \sin \emptyset) \\ & + (\{(\%\text{Reactance} \times \cos \emptyset) - (\%\ \text{Resistance} \times \sin \emptyset)\}^2/200\end{aligned} \tag{8.50}$$

## 8.15 Tank Calculation

$$\text{Tank Breath} - B = ((\text{HV coil OD} + 256)/5) * 5 - 15 \tag{8.51}$$

$$\text{Tank Length} - L = ((\text{Core CL} * 2 + \text{Tank B})/5) * 5 + 30 \tag{8.52}$$

$$\text{Tank Height} - H = ((\text{WH} * 2 + \text{Core Dia}) + 150)/5)^* 5 + 75 \tag{8.53}$$

## 8.16 Pressure and Vacuum Calculation

(1) Tank Height in cm
(2) Tank to Conservator 1/2 Level in cm.

Total $(T) = (1 + 2) =$ ___cm.

$$\begin{aligned}\text{Head Pressure}(P) &= 0.00089 \times \text{Head of Oil}(T)(\text{cm}) \\ &= 0.00089 \times (T) = 0.25\ \text{kg/cm}^2. \\ &= \text{Kg/cm}^2/\text{PSI}\end{aligned}$$

The required pressure for the pressure test is $P + 5$ PSI.

## 8.17 Oil Calculation

| Material | Weight | Density | Displacement weight/Density |
|---|---|---|---|
| Copper | Weight | Density | Displacement |
| CRGO | Weight | Density | Displacement |
| MS frame | Weight | Density | Displacement |
| Connection copper | Weight | Density | Displacement |
| Insulation | Weight | Density | Displacement |
| | Total: | | Displacement |
| Total Volume of Tank = $L \times B \times H$ cm$^3$ | | | |
| Total Volume of oil ($A$): Total Volume of Tank − Total Displacement | | | |
| Oil in Conservator ($B$): | 10% of Total Oil Volume | | |
| Radiator Oil ($C$): | (a) Ltrs in one finns | | |
| | (b) No. of fins in one radiator | | |
| | (c) Total number of radiators | | |
| Total oil in radiator = (a*b*c) | | | |
| Total Oil Qty = A + B + C | | | |

### 8.17.1 Radiator and Cooling Calculation

| | |
|---|---|
| CC = | (tank height-100–1.1*core dia)/100)100–400 |
| Nos section | ((radiator dissipation/(W/section)/(Nos of rad + 1)) + 3 |
| Nos radiators | (Tank Length/520 + 1)*2 |
| Rad dissipation | (NLL + LL)*(1 + positive tap var/200)-tank dissipate |
| Tank dissipation | Tank surface area*6*Top oil rise |
| Tank surface area | (Tank length + tank breath)*tank height/1000000 |
| A | (Oil rise -30)/5 |
| Watt Diss (Q) | (Oil rise temperature factor @ CC for table |
| B | (Oil rise/(A*5 + 30) |
| C (VF const.) | ((Tank ht-100-CC/2)-(WH/2 + Core Dia * 1.2)) |
| VF | Concerning the value of C take VF from the table |
| D (HF const.) | Tank ht/ (Nos of radiator/2–1) |
| HF | Concerning the value of D Take VF from the table |
| E (SF const) | Radiator dissip/A/B/VF/HF |
| SF | Concerning the value of E/F take SF from the table |

(continued)

(continued)

| | |
|---|---|
| F (WF const) | Concerning the value of E Take WF from the table |
| W/section | Q*B.$V_F$.$H_F$.$S_F$ |

## 8.18 Computer-Based Power Transformer Design

Based on the above calculations and standard technical data as per Appendix (C), the power transformer design is formulated in MS Excel. All the parameters, like losses, % Z, winding geometries, and core geometries, temperature rise, etc., are acquired with the help of mathematical formulation concerning input technical data like rating, HV & LV voltages, tap range, vector group, etc. The design module is shown in Fig. 8.3, and the respective output is shown in Table 8.2.

The computer-based program of transformer design was executed several times to mitigate the technical parameters like NLL, load loss, and %Z. The result summary of some designs is presented in Table 8.2.

## 8.19 MATLAB-Based Program for Transformer Design Optimization

It is difficult to decide whether the conventional techniques for designing the transformer are optimal designing or not. To define the optimum designing parameters with particular objectives including some specific design constraints it is required to develop some program in MATLAB tool [26, 27]. The whole design consideration and design procedure are given in Fig. 8.4. The pattern of the MATLAB TDO (transformer design optimization) program is shown in Fig. 8.5a.

Following the "Iterative Method" of the TDO program execution, 19 outcomes are achieved by meeting the constraints. These results are displayed in Fig. 8.5b. Tables 8.3 and 8.4, respectively, show the list of variables and constrained considered for the TDO program. The outcome of this iterative TDO method in the form of a graphical representation is shown in Fig. 8.6, and it includes flux density versus NLL, current density versus load loss, *K*-factor versus NLL, etc.

In this method, the above results are saved in an Excel sheet. So, from that sheet, we can pick any suitable design as per customer requirement and design the transformer.

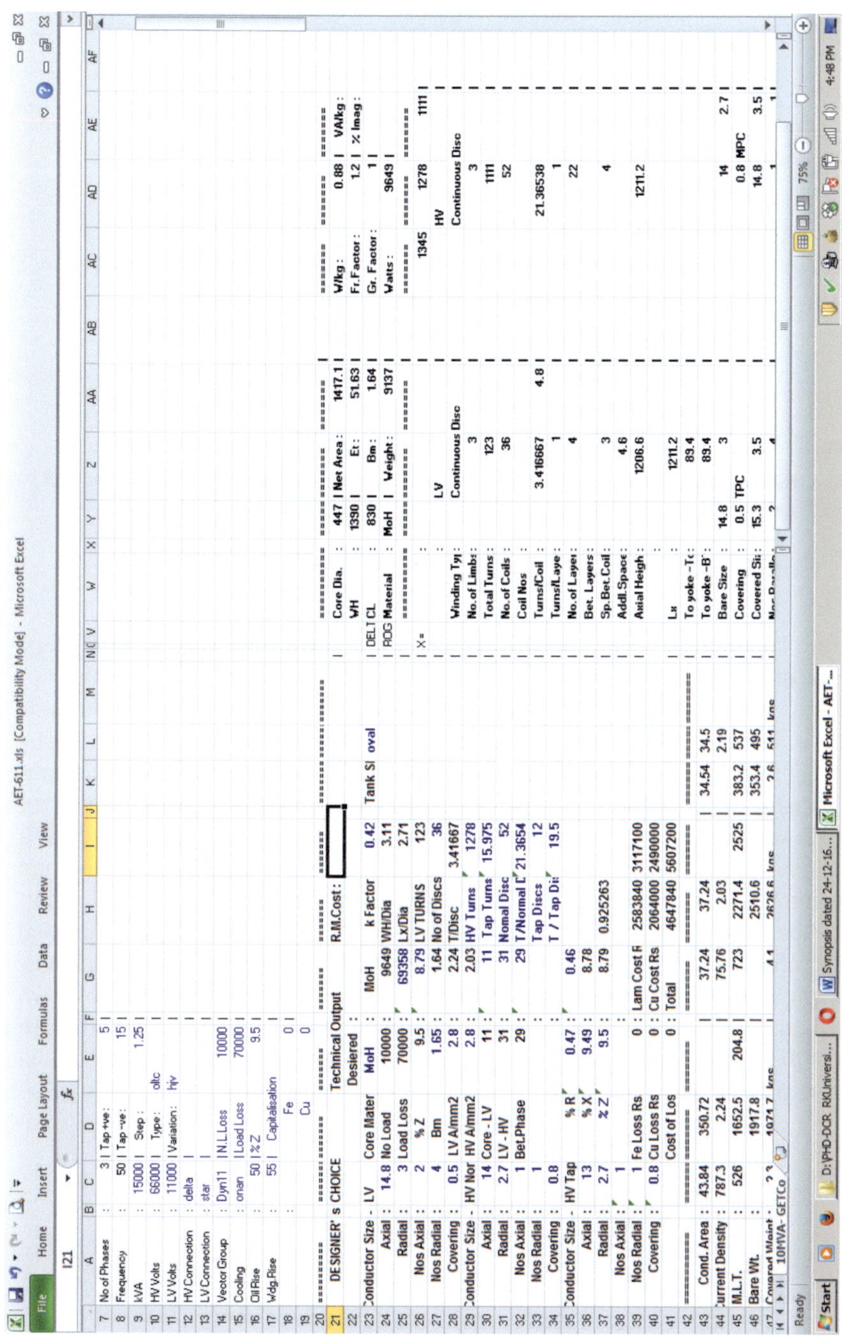


**Fig. 8.3** Programming for design of power transformer using MS-excel

**Table 8.2** Design data of power transformer conventional design program by Ms-excel tool

| Parameters | Result 1 | Result 2 | Result 3 | Result 4 | Result 5 | Result 6 |
|---|---|---|---|---|---|---|
| k-factor | 0.42 | 0.417 | 0.428 | 0.432 | 0.441 | 0.432 |
| $B_m$ (wb/m$^2$) | 1.60 | 1.65 | 1.58 | 1.57 | 1.59 | 1.6 |
| Del-LV (A/mm$^2$) | 2.23 | 2.39 | 2.51 | 2.48 | 2.45 | 2.60 |
| Del-HV (A/mm$^2$) | 2.81 | 2.62 | 2.67 | 2.58 | 2.47 | 2.65 |
| No-load losses (kW) | 9.92 | 9.79 | 9.91 | 10.01 | 9.88 | 9.73 |
| Load losses (kW) | 66.79 | 66.85 | 67.91 | 68.14 | 67.66 | 67.42 |
| % Z | 8.81 | 8.72 | 8.95 | 9.43 | 9.07 | 8.90 |
| % $\eta$ | 99.13 | 99.22 | 99.10 | 99.21 | 99.20 | 99.14 |
| GLV | 9.56 | 9.25 | 9.76 | 9.47 | 10.03 | 9.94 |
| GHV | 20.87 | 21.92 | 21.59 | 22.06 | 20.88 | 20.73 |
| Wcore (kg) | 11,133.4 | 10,960.5 | 10,750.5 | 11,077.8 | 10,990.8 | 11,170.6 |
| Wcu (kg) | 5026 | 5250 | 5180.9 | 5370.8 | 5266.9 | 5305.6 |
| Total cost (Rs.) | 6,305,692 | 6,330,566 | 6,405,511 | 6,290,520 | 6,414,953 | 6,348,652 |

## 8.20 Conclusion

The entire chapter gives full information about all transformer design parameters, with basic information. It also gives a computerized design sequence in the manner of a flowchart with the use of national and international standards and the industrial code of practice. The computer program is developed based on an experiment related to the prevailing design methodology. Later, the MATLAB software is used to get the optimized transformer design parameters and the design solution. As a perfect designer, it presents the analytical and graphical comparison between various parameters. It also includes all the parameters for the transformer design, including oil, pressure and vacuum, efficiency and regulations, radial bursting force inrush, dynamic short circuit, capacitance, AT, winding gradient, core steps calculations, etc., are included in depth. While using a MATLAB-based algorithm for transformer design, various constraints and variables are considered as per standard design practices. Initially, the calculations are done in Excel as per the set of equations derived for the effective design of the transformer. Several design parameters are obtained by a conventional design program in an Excel sheet.

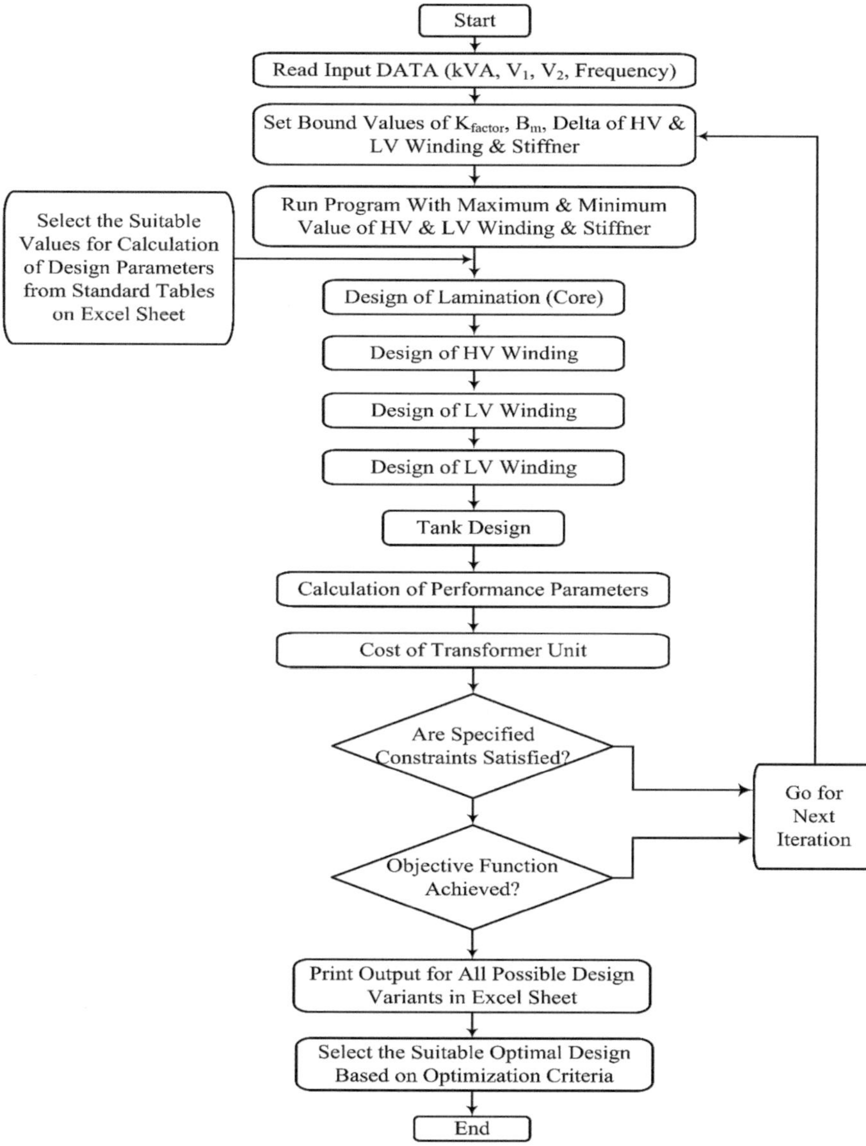


**Fig. 8.4** MATLAB flowchart for transformer design optimization process using iterative method

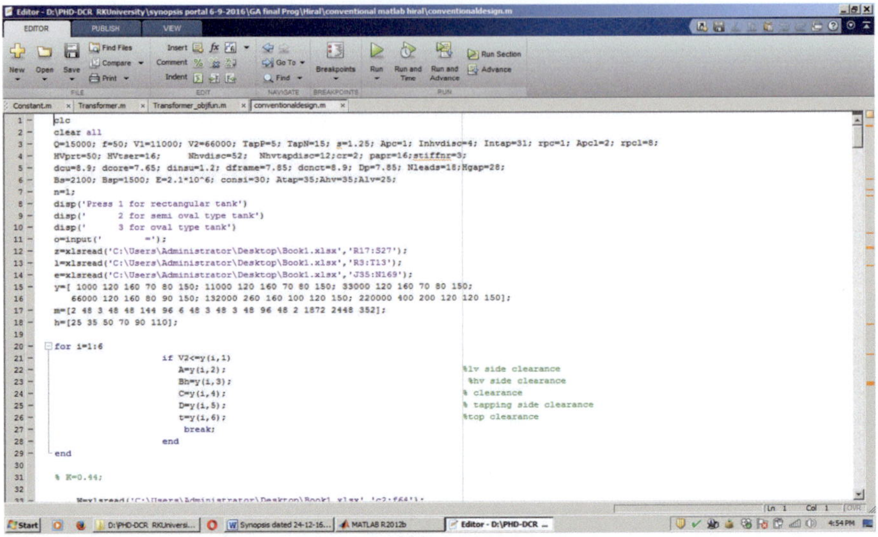


(A)

| 1 | K | Bm | delhv | dellv | stiffnr | L | B | H | Closs | Ldloss | Z | effi | GL | GH | dwss | Voil | Tocost |
|---|---|---|---|---|---|---|---|---|---|---|---|---|---|---|---|---|---|
| 2 | 0.45 | 1.46 | 2.6 | 2.5 | 2 | 2663.24 | 1091.08 | 2006.094 | 9092.529 | 67607.69 | 8.688832 | 99.22538 | 9.024926 | 22.79127 | 8.203506 | 5.83E+09 | 6417792 |
| 3 | 0.45 | 1.47 | 2.6 | 2.4 | 2 | 2669.662 | 1093.221 | 2004.709 | 9152.108 | 66404.32 | 8.782509 | 99.23906 | 9.21247 | 22.37667 | 8.922365 | 5.85E+09 | 6474726 |
| 4 | 0.45 | 1.47 | 2.6 | 2.5 | 2 | 2658.202 | 1089.401 | 2000.889 | 9134.952 | 67416.65 | 8.667817 | 99.22755 | 9.026163 | 22.79441 | 8.087925 | 5.79E+09 | 6385213 |
| 5 | 0.45 | 1.48 | 2.6 | 2.4 | 2 | 2664.676 | 1091.559 | 1999.556 | 9193.886 | 66219.21 | 8.761528 | 99.24117 | 9.213796 | 22.3799 | 8.058549 | 5.82E+09 | 6431357 |
| 6 | 0.45 | 1.48 | 2.6 | 2.5 | 2 | 2653.215 | 1087.738 | 1995.736 | 9176.631 | 67227.55 | 8.647015 | 99.2297 | 9.027395 | 22.79754 | 7.974835 | 5.76E+09 | 6353074 |
| 7 | 0.45 | 1.49 | 2.6 | 2.4 | 2 | 2659.739 | 1089.913 | 1994.455 | 9348.947 | 66035.97 | 8.740758 | 99.24325 | 9.215117 | 22.38313 | 7.946927 | 5.78E+09 | 6385184 |
| 8 | 0.45 | 1.49 | 2.6 | 2.5 | 2 | 2648.279 | 1086.093 | 1990.635 | 9331.379 | 67040.36 | 8.626423 | 99.23183 | 9.028622 | 22.80065 | 7.864166 | 5.73E+09 | 6307203 |
| 9 | 0.45 | 1.5 | 2.6 | 2.4 | 2 | 2654.852 | 1088.284 | 1989.405 | 9388.385 | 65854.57 | 8.720196 | 99.24532 | 9.216431 | 22.38633 | 7.837676 | 5.75E+09 | 6353695 |
| 10 | 0.45 | 1.5 | 2.6 | 2.5 | 2 | 2643.392 | 1084.464 | 1985.585 | 9370.722 | 66855.04 | 8.606037 | 99.23394 | 9.029843 | 22.80375 | 7.75585 | 5.69E+09 | 6275948 |
| 11 | 0.45 | 1.51 | 2.6 | 2.4 | 2 | 2650.014 | 1086.671 | 1984.406 | 9539.369 | 65674.97 | 8.699839 | 99.24736 | 9.21774 | 22.38953 | 7.730728 | 5.71E+09 | 6322622 |
| 12 | 0.45 | 1.51 | 2.6 | 2.5 | 2 | 2638.554 | 1082.851 | 1980.586 | 9521.4 | 66671.57 | 8.585854 | 99.23602 | 9.031059 | 22.80683 | 8.351614 | 5.66E+09 | 6255962 |
| 13 | 0.45 | 1.52 | 2.6 | 2.4 | 2 | 2645.224 | 1085.075 | 1979.455 | 9576.586 | 65497.14 | 8.679683 | 99.24938 | 9.219044 | 22.39271 | 8.325629 | 5.68E+09 | 6302806 |
| 14 | 0.45 | 1.52 | 2.6 | 2.5 | 2 | 2633.763 | 1081.254 | 1975.635 | 9558.525 | 66489.91 | 8.565871 | 99.23809 | 9.03227 | 22.8099 | 8.238282 | 5.63E+09 | 6225499 |
| 15 | 0.45 | 1.53 | 2.6 | 2.4 | 2 | 2640.48 | 1083.493 | 1974.554 | 9834.157 | 65321.06 | 8.659725 | 99.25138 | 9.220342 | 22.39587 | 8.213691 | 5.65E+09 | 6272514 |
| 16 | 0.45 | 1.53 | 2.6 | 2.5 | 2 | 2629.02 | 1079.673 | 1970.734 | 9815.588 | 66310.03 | 8.546083 | 99.24014 | 9.033476 | 22.81296 | 8.127309 | 5.59E+09 | 6195433 |
| 17 | 0.45 | 1.54 | 2.6 | 2.4 | 2 | 2635.783 | 1081.928 | 1969.7 | 9868.417 | 65146.7 | 8.639962 | 99.25337 | 9.221635 | 22.39903 | 8.104064 | 5.62E+09 | 6242614 |
| 18 | 0.45 | 1.54 | 2.6 | 2.5 | 2 | 2624.323 | 1078.108 | 1965.88 | 9849.761 | 66131.91 | 8.526489 | 99.24216 | 9.034677 | 22.81601 | 8.018629 | 5.56E+09 | 6165758 |
| 19 | 0.45 | 1.55 | 2.6 | 2.5 | 2 | 2619.671 | 1076.557 | 1961.073 | 9991.957 | 65955.52 | 8.507085 | 99.24417 | 9.035873 | 22.81904 | 7.91218 | 5.53E+09 | 6136465 |

(B)

**Fig. 8.5** **a** Programming for design of power transformer using MATLAB and **b** respective result using iterative method

**Table 8.3** List of variables for the TDO program

| Sr. No. | Variable | Range |
|---|---|---|
| 1 | *K*-factor | 0.35 to 0.55 |
| 2 | Flux density (Bm) | 1.45 to 1.75 $Wb/m^2$ |
| 3 | Current density of LV winding | 1.9 to 3.5 $A/mm^2$ |
| 4 | Current density of HV winding | 1.9 to 3.5 $A/mm^2$ |
| 5 | No. of stiffener | 2 to 8 |

**Table 8.4** List of design constraints for the TDO program

| Sr. No. | Constraints | Range |
|---|---|---|
| 1 | No-load loss | < 11,000 W |
| 2 | Load loss | < 70,000 W |
| 3 | Percentage impedance | 7 < %z < 10 |
| 4 | Efficiency | > 99% |
| 5 | The gradient of LV winding | 9 < GLV < 23 |
| 6 | The gradient of HV winding | 9 < GHV < 23 |
| 7 | Deflection | 5 < def < 9 mm |

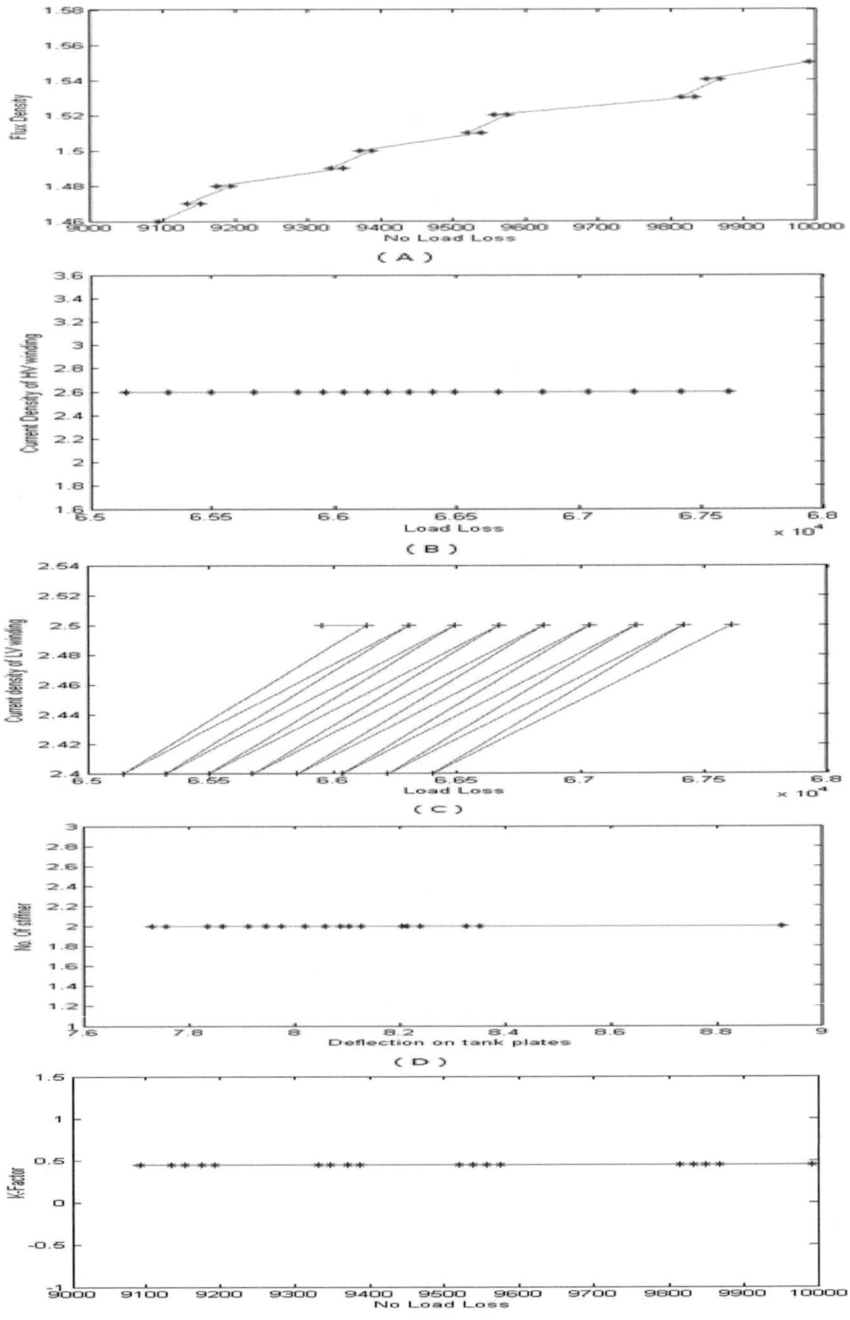


**Fig. 8.6** Graphical representation of iterative method TDO, **a** flux density versus NLL, **b** current density (HV) versus load loss, **c** current density (LV) versus load loss, **d** stiffener versus deflection in tank, and **e** *K*-factor versus no load loss

# References


1. Chothani N, Raichura M, Patel D (2023) Transformer infrastructure for power grid. In: Advancement in power transformer infrastructure and digital protection. Springer Nature Singapore, Singapore, pp 1–26. https://doi.org/10.1007/978-981-99-3870-4_1
2. Nilesh Chothani DP, Raichura M (2023) Advancement in power transformer infrastructure and digital protection, 1st edn. Studies in infrastructure and control, Springer, Singapore
3. Patel D, Chothani N (2020) Digital protective schemes for power transformer, first. In: Power systems. Springer Singapore, Singapore. [Online]. Available: https://books.google.co.in/books?id=tjaSzQEACAAJ
4. Zhou X et al (2023) A design method of partially interleaved winding structure with low leakage inductance for planar transformer application. IEEE Trans Power Electron 38(5):6366–6379. https://doi.org/10.1109/TPEL.2023.3242109
5. Pan Y, Geng J, Zhou L, Lai L, Zhan G, Shao H (2024) Multifunctional wideband digital sampling system for the calibration of conventional transformers and electronic transformers. IEEE Trans Instrum Meas 73:1–8. https://doi.org/10.1109/TIM.2024.3420370
6. Patel D, Desai A, Chothani N (2023) Fixed frequency current control-based boost rectifier analysis for single phase solid state transformer. In: 2023 international conference for advancement in technology (ICONAT), pp 1–6. https://doi.org/10.1109/ICONAT57137.2023.10080301
7. Chothani N, Patel DD, Desai A (2023) Hysteresis current control-based boost rectifier analysis for single phase solid state transformer. In: 2023 international conference on recent advances in electrical, electronics and digital healthcare technologies (REEDCON), pp 385–390. https://doi.org/10.1109/REEDCON57544.2023.10151262
8. Patel DD, Chothani N, Parmar BD (2023) Design and development of solid state transformer with different current control techniques. In: 2023 IEEE 11th region 10 humanitarian technology conference (R10-HTC), pp 414–419. https://doi.org/10.1109/R10-HTC57504.2023.10461788
9. Guo Z et al (2023) A novel high insulation 100 kW medium frequency transformer. IEEE Trans Power Electron 38(1):112–117. https://doi.org/10.1109/TPEL.2022.3205646
10. Chakraborty SS, Dey S, Hatua K (2023) Design of a three-winding transformer for power decoupling of a three-port series resonant converter for an integrated on-board EV charger. IEEE Trans Power Electron 38(11):14262–14273. https://doi.org/10.1109/TPEL.2023.3308776
11. Wang H, Zhang L, Sun Y, Zou L (2024) Research on the influence mechanism of harmonic components on the noise distribution characteristics of converter transformers. Int J Electr Power Energy Syst 160:110095. https://doi.org/10.1016/J.IJEPES.2024.110095
12. Mao B et al (2024) Designing ultra-broadband terahertz polarization converters based on the transformer model. Opt Commun 559:130434. https://doi.org/10.1016/J.OPTCOM.2024.130434
13. Hashemi MH, Kiliç U, Dikmen S (2023) Applications of novel heuristic algorithms in design optimization of energy-efficient distribution transformer. IEEE Access 11:15968–15980. https://doi.org/10.1109/ACCESS.2023.3245327
14. Lee GS, Kim S, Bae S (2023) Efficient design method for a forward-converter transformer based on a KNN–GRU–DNN model. IEEE Trans Power Electron 38(1):73–78. https://doi.org/10.1109/TPEL.2022.3203480
15. Mahamed M, Seyyedbarzegar S (2024) Design and thermal analysis of a 250 MVA HTS transformer for substation of offshore wind farms. Phys C Supercond Appl 620:1354503. https://doi.org/10.1016/J.PHYSC.2024.1354503
16. Dawood K, Tursun S (2024) Optimized gap positions for improved leakage impedance in dry-type transformer design. Results Eng 23:102632. https://doi.org/10.1016/j.rineng.2024.102632
17. Tarafdar-Hagh M, Hasanlouei N, Muttaqi K, Taghizad-Tavana K, Nojavan S, Abdi M (2024) Application of saturated core transformer as inrush current limiter. e-Prime—Adv Electr Eng Electron Energy 9:100677. https://doi.org/10.1016/J.PRIME.2024.100677

18. Subramanya K, Chelliah TR (2024) DC bias impact analysis on the capability of power transformer and failure risks. Eng Fail Anal 163:108537. https://doi.org/10.1016/J.ENGFAILANAL.2024.108537
19. Patel D, Chothani N (2020) Real-time monitoring and adaptive protection of power transformer. In: Digital protective schemes for power transformer, Singapore: Springer Singapore, pp 173–190. https://doi.org/10.1007/978-981-15-6763-6_7
20. Raichura M, Chothani NG, Patel D (2019) Real-time monitoring and adaptive protection of power transformer to enhance smart grid reliability. Electr Control Commun Eng 15(2):104–112, https://doi.org/10.2478/ecce-2019-0014
21. Raichura M, Chothani N, Patel D (2021) Review of methodologies used for detection of magnetising inrush and fault conditions in power transformer. IET Energy Syst Integr 3:109–129. https://doi.org/10.1049/esi2.12012
22. Raichura M, Chothani N, Patel D, Sharma J (2020) Methodologies for the detection of magnetizing inrush and fault condition in power transformer. In: 2020 IEEE international conference on computing, power and communication technologies (GUCON), pp 146–151. https://doi.org/10.1109/GUCON48875.2020.9231065
23. Amoiralis EI, Tsili MA, Kladas AG (2009) Transformer design and optimization: a literature survey. IEEE Trans Power Deliv 24(4):1999–2024. https://doi.org/10.1109/TPWRD.2009.2028763
24. Harlow JH (ed) (2012) Electric power transformer engineering, 3rd edn. CRC press, Tailor and Francies group. https://doi.org/10.1201/b12110
25. Georgilakis PS (2009) Spotlight on modern transformer design, 1st edn. Springer London. https://doi.org/10.1007/978-1-84882-667-0
26. Heathcote MJ (1998) The J & P transformer book, p 957
27. Pratap R (2002) Getting started with MATLAB—a quick introduction for scientists and engineers, vol 8

MIX
Papier aus verantwortungsvollen Quellen
Paper from responsible sources
FSC® C105338

If you have any concerns about our products,
you can contact us on
**ProductSafety@springernature.com**

In case Publisher is established outside the EU,
the EU authorized representative is:
**Springer Nature Customer Service Center GmbH**
**Europaplatz 3, 69115 Heidelberg, Germany**

Printed by Libri Plureos GmbH
in Hamburg, Germany